D0163269

REVERSE MATHEMATICS

REVERSE MATHEMATICS

PROOFS FROM THE INSIDE OUT

■■■■■

John Stillwell

PRINCETON UNIVERSITY PRESS
PRINCETON AND OXFORD

Published by Princeton University Press, 41 William Street, Princeton, New Jersey 08540
In the United Kingdom: Princeton University Press, 6 Oxford Street, Woodstock, Oxfordshire OX20 1TR

press.princeton.edu

Jacket images courtesy of Alamy

Library of Congress Cataloging-in-Publication Data
Names: Stillwell, John, author.
Title: Reverse mathematics : proofs from the inside out / John Stillwell.
Description: Princeton : Princeton University Press, [2018] | Includes bibliographical references and index.
Identifiers: LCCN 2017025264 | ISBN 9780691177175 (hardback)
Subjects: LCSH: Reverse mathematics. | BISAC: MATHEMATICS / History & Philosophy. | MATHEMATICS / General. | MATHEMATICS / Logic. | SCIENCE / History.
Classification: LCC QA9.25 .S75 2018 | DDC 511.3–dc23 LC record available at https://lccn.loc.gov/2017025264

British Library Cataloging-in-Publication Data is available

This book has been composed in Minion Pro

Printed on acid-free paper. ∞

Printed in the United States of America
10 9 8 7 6 5 4 3 2 1

To Elaine

■■■■■

Contents

Preface

This is a book about the foundations of mathematics—a topic once of interest to outstanding mathematicians, such as Dedekind, Poincaré, and Hilbert, but today sadly neglected. This neglect is unfortunate for several reasons:

- As mathematics splits into more and more specialties, the need for a unifying viewpoint becomes more acute.
- Foundations unify not only mathematics but also the neighboring disciplines of computer science and physics.
- Recent advances in mathematical logic throw new light on the foundations of analysis, and on the elusive concept of mathematical "depth."

This book aims at the last point in particular, by focusing on the topic of *reverse mathematics*.

As its name suggests, reverse mathematics looks at the concept of proof in the opposite to normal direction. Instead of seeking the consequences of given axioms, it seeks the axioms needed to prove given theorems. This is actually an old idea, at least in the foundations of geometry. From the time of Euclid until the nineteenth century it was a burning question whether the parallel axiom was needed to prove theorems such as the Pythagorean theorem. We review the history of the parallel axiom in chapter 1 of this book, as a case study in reverse mathematical ideas, together with the similar story of the axiom of choice in set theory.

Although both these axioms illustrate the idea of reverse mathematics, the subject as it is understood today lies mostly in a narrow but important region *between* geometry and set theory: the theory of real numbers, which is the foundation of calculus, analysis, and most of mathematical physics. (Reverse mathematics has also made interesting contributions to algebra, combinatorics, and topology which we mention more briefly.)

The real numbers, as we understand them today, emerged from nineteenth century efforts to *arithmetize* analysis and geometry. By building real numbers from sets of rational numbers (and hence, ultimately, from sets of natural numbers) it becomes possible to encode sequences of real numbers and arbitrary continuous functions—and hence most of the objects of analysis—by sets of natural numbers. We review the arithmetization of analysis, and also the basic theorems of analysis, in chapters 2 and 3. After this we are ready to ask: which *axioms* do we need to prove these basic theorems? The answer, roughly, is a set of axioms for the natural numbers (the *Peano axioms*) plus a suitable *set existence axiom*.

Now set existence axioms come in various *strengths*, depending on the strength of the theorems we wish to prove. The lowest useful strength turns out to be intimately related to the foundations of *computation*: it asserts the existence of computable sets. This in turn involves a study of the concept of computation, which merges with analysis because both have a common basis in arithmetic. After an informal introduction to computability in chapter 4 we develop a formal concept of computation, and its arithmetization, in chapter 5.

In chapters 6 and 7 we bring together the ideas of analysis, arithmetic, and computation in some axiom systems for analysis, known as RCA_0, WKL_0, and ACA_0. These systems, which are distinguished mainly by set existence axioms of increasing strength, between them prove most of the basic theorems of analysis. More remarkably, they sort the basic theorems into three levels because, once above the "base" level of RCA_0, most theorems are *equivalent* to the set existence axiom of the system that proves them. This makes each of these set existence axioms the "right axiom" in the sense articulated by Friedman (1975):

> When a theorem is proved from the right axioms, the axioms can be proved from the theorem.

We will see, for example, that RCA_0 can prove the intermediate value theorem; the defining axiom of WKL_0 is the right axiom to prove the Heine-Borel theorem and the extreme value theorem; and the defining axiom of ACA_0 is the right axiom to prove the Cauchy convergence criterion and the Bolzano-Weierstrass theorem.

Thus in reverse mathematics we meet the usual cast of characters from an introductory real analysis course, but in an entirely new story.

In chapter 8 we give some glimpses of the bigger picture of analysis, computation, and logic, which will hopefully prepare the reader for specialist treatments of reverse mathematics, notably Simpson (2009). The present book is very much for non-specialists—in some ways a sequel to my book *Elements of Mathematics. From Euclid to Gödel.* It develops computability and logic far enough to explain results that *Elements of Mathematics* could only mention, but the latter book is not a prerequisite for this one. Anyone at an upper undergraduate level with an interest in foundations should be able to approach the reverse mathematics in this book directly. The same goes, of course, for professional mathematicians who want to refresh their memory of foundations and to see how the subject has reinvented itself in recent times.

Acknowledgements. I thank Harvey Friedman for information on the history of reverse mathematics, Keita Yokoyama for his insights on topology, and two anonymous referees for many helpful comments and corrections. My wife Elaine as usual did sterling work in proofreading, and Vickie Kearn and her team at Princeton University Press were ever-helpful and meticulous in the production of the book.

John Stillwell
San Francisco, 24 November 2016

REVERSE MATHEMATICS

CHAPTER 1

Historical Introduction

The purpose of this introductory chapter is to prepare the reader's mind for *reverse mathematics*. As its name suggests, reverse mathematics seeks not theorems but the right axioms to prove theorems already known. The criterion for an axiom to be "right" was expressed by Friedman (1975) as follows:

> When the theorem is proved from the right axioms, the axioms can be proved from the theorem.

Reverse mathematics began as a technical field of mathematical logic, but its main ideas have precedents in the ancient field of geometry and the early twentieth-century field of set theory.

In geometry, the parallel axiom is the right axiom to prove many theorems of Euclidean geometry, such as the Pythagorean theorem. To see why, we need to separate the parallel axiom from the *base theory* of Euclid's other axioms, and show that the parallel axiom is not a theorem of the base theory. This was not achieved until 1868. It is easier to see that the base theory can prove the parallel axiom *equivalent* to many other theorems, including the Pythagorean theorem. This is the hallmark of a good base theory: what it cannot prove outright it can prove equivalent to the "right axioms."

Set theory offers a more modern example: a base theory called ZF, a theorem that ZF cannot prove (the well-ordering theorem) and the "right axiom" for proving it—the axiom of choice.

From these and similar examples we can guess at a base theory for analysis, and the "right axioms" for proving some of its well-known theorems.

1.1 EUCLID AND THE PARALLEL AXIOM

The search for the "right axioms" for mathematics began with Euclid, around 300 BCE, when he proposed axioms for what we now call *Euclidean geometry*. Euclid's axioms are now known to be incomplete; nevertheless, they outline a complete system, and they distinguish between really obvious "basic" axioms and a less obvious one that is crucial for obtaining the most important theorems. For historical commentary on the axioms, see Heath (1956).

The basic axioms say, for example, that there is a unique line through two distinct points and that lines are unbounded in length. Also basic, though expressed only vaguely by Euclid, are criteria for *congruence of triangles*, such as what we call the "side angle side" or SAS criterion: if two triangles agree in two sides and the included angle then they agree in all sides and all angles. (Likewise ASA: they agree if they agree in two angles and the side between them.)

Using the basic axioms it is possible to prove many theorems of a rather unsurprising kind. An example is the *isosceles triangle theorem*: if a triangle ABC has side AB = side AC then the angles at B and C are equal. However, the basic axioms fail to prove the signature theorem of Euclidean geometry, the *Pythagorean theorem*, illustrated by figure 1.1.

Figure 1.1 : The Pythagorean theorem

As everybody knows, the theorem says that the gray square is equal to the sum of the black squares, but the basic axioms cannot even prove the *existence* of squares. To prove the Pythagorean theorem, as Euclid realized, we need an axiom about infinity: the *parallel axiom*.

The Parallel Axiom

I call the parallel axiom an axiom about infinity because it is about lines that do not meet, *no matter how far they are extended*—and one of Euclid's basic axioms is that lines can be extended indefinitely. Thus parallelism cannot be "seen" unless we have the power to see to infinity, and Euclid preferred not to assume such a superhuman power. Instead, he gave a criterion for lines *not* to be parallel, since a meeting of lines can be "seen" a finite distance away.

Parallel axiom. If a line n falling on lines l and m (figure 1.2) makes angles α and β with $\alpha + \beta$ less than two right angles, then l and m meet on the side on which α and β occur.

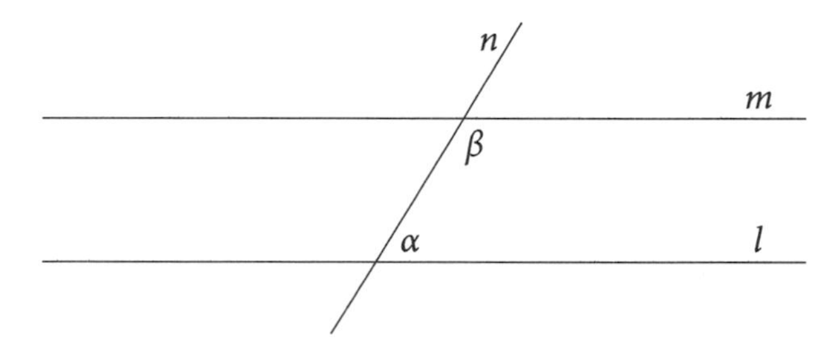

Figure 1.2 : Angles involved in the parallel axiom

It follows that if $\alpha + \beta$ equals two right angles (that is, a straight angle) then l and m do *not* meet. Because if they meet on one side (forming a triangle) they must meet on the other (forming a congruent triangle, by ASA), since there are angles α and β on both sides and one side in common (figure 1.3). This contradicts uniqueness of the line through any two points.

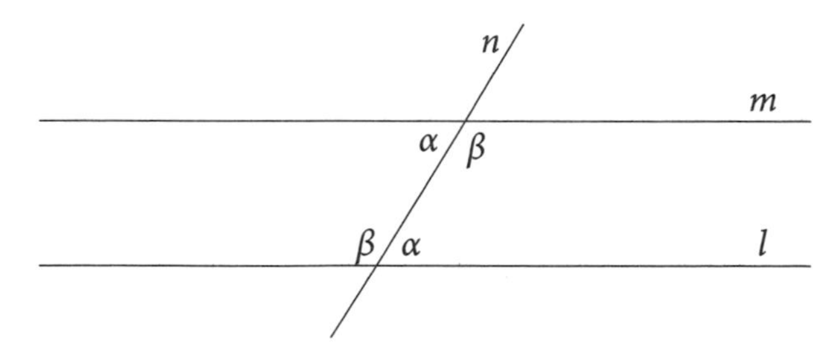

Figure 1.3 : Parallel lines

Thus Euclid's axiom about *non*-parallel lines implies that parallel lines exist. From parallel lines we quickly get the theorem that the angle sum of a triangle is a straight angle (or π, as we will write it from now on), by the construction shown in figure 1.4. From this we find in turn that an isosceles triangle with angle $\pi/2$ between its equal sides has its other angles equal to $\pi/4$, so putting two such triangles together makes a square.

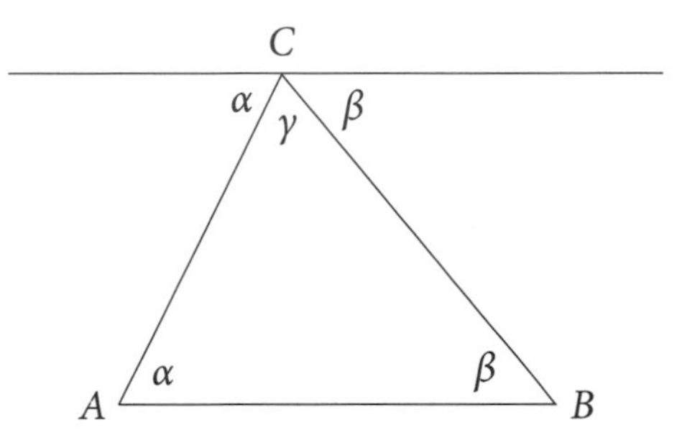

Figure 1.4 : Angle sum of a triangle

The proof of the Pythagorean theorem can now get off the ground, and there are many ways to complete it. Probably the one most easily "seen" is shown in figure 1.5, in which the gray square and the two black squares both equal the big square minus four copies of the right-angled triangle.

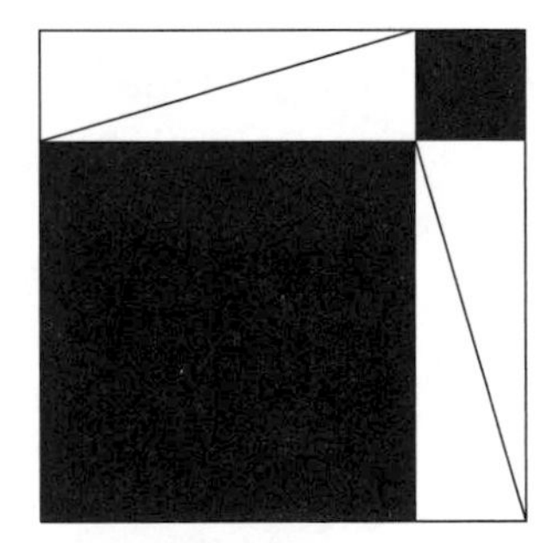

Figure 1.5 : Seeing the Pythagorean theorem

Equivalents of the Parallel Axiom

Many mathematicians considered the parallel axiom to be a "blemish" on Euclid's system—this is precisely what Saccheri (1733) called it—so they tried to show that it followed from the other axioms. Their attempts usually took the form of deducing the parallel axiom from a seemingly more obvious statement, in the hope of reducing the problem to a simpler one. Some of the statements found to imply the parallel axiom were:

- existence of rectangles (al-Haytham, al-Tusi in medieval times),
- existence of similar triangles of different sizes (Wallis in 1693),
- angle sum of triangle = π (in Legendre's *Éléments de géométrie*, 1823),
- three noncollinear points lie on a circle (Farkas Bolyai (1832)).

All of these theorems follow from the parallel axiom, so they are equivalent to it in *strength*, in the sense that their equivalence to the parallel axiom can be proved using only the other axioms. Of course, this notion of equivalent strength is trivial if the parallel axiom itself is provable from the other axioms, but by 1830 the hopes of such a proof were fading. Farkas Bolyai's own son, János, was one of the main explorers of a hypothetical *non*-Euclidean geometry in which the parallel axiom (and hence the four theorems above) is *false*, yet Euclid's other axioms are true.

But before seeing non-Euclidean geometry, it helps to look at geometry on the sphere. Spherical geometry is clearly different from the Euclidean geometry of the plane—not only in the absence of parallels, but also in the absence of infinite lines—yet they share a common language of "points," "lines," and "angles." Seeing two different interpretations of these words will make it easier to grasp yet another interpretation, or *model*—a model of non-Euclidean geometry.

1.2 SPHERICAL AND NON-EUCLIDEAN GEOMETRY

Just as circles and lines in the plane are part of two-dimensional Euclidean geometry, spheres and planes are part of *three*-dimensional Euclidean geometry. Indeed they are mentioned, though not deeply studied, in Euclid's *Elements*, Book XI. The ancient Greeks made a serious study of spherical geometry, particularly spherical trigonometry, in their study of astronomy, because the stars appear from the earth to be fixed on a heavenly sphere. Later, navigators on the earth also took an interest in spherical geometry. For them, the natural concept of "line" is that of a *great circle*—the intersection of the sphere with a plane through its center—because a great circle gives the shortest distance between any two of its points. The concept of "angle" between any two such "lines" also makes sense, as the angle between the corresponding planes (or, what comes to the same thing, the angle between the tangents to the great circles).

Indeed, it is often easier to describe a spherical triangle by its angles rather than the lengths of its sides. All spherical triangles with the same angles in fact have the same size, because of a famous theorem of Harriot[1] from 1603: *the angle sum of a spherical triangle, minus* π, *is proportional to its area*. There are several ways to tile the surface of the sphere with congruent triangles. Figure 1.6 shows one in which the sphere is divided into 48 triangles, each of which has angles $\pi/2, \pi/3, \pi/4$. Alternate triangles have been cut out of the sphere, to make it easier to see them all, and the sphere has been illuminated from the inside. This then is the standard

Figure 1.6 : Tiling the sphere with triangles

model of spherical geometry: "points" are ordinary points on the sphere, "lines" are great circles, and "angles" are the angles between the tangents to the great circles at their point of intersection. "Distance," if we wish to use the concept, is the distance between points on the sphere, measured along the (shorter) piece of the great circle connecting them.

Now we move to another model, by *projecting the sphere onto the plane*. Specifically, we use the light inside the sphere (at its north pole) to cast a shadow on the plane. The result is shown in figure 1.7. The pic-

[1]Thomas Harriot was mathematical consultant to Sir Walter Raleigh, and traveled with him on some of his voyages.

Figure 1.7 : Projecting the sphere onto the plane

ture shows two remarkable features of projection from the north pole, which is known as *stereographic* projection:

- circles map to circles (or, in exceptional cases, to straight lines, which we might call "circles of infinite radius"), and
- angles are preserved.

Thus "points" are still points, "lines" are still circles, and "angle" is still the angle between the tangents to the circles. "Distance," alas, is not a Euclidean distance of any kind, since equal distances on the sphere can be mapped to unequal Euclidean distances in the plane. Likewise, "area" is not Euclidean area, but we can easily measure it by the angle sum minus π.

Strictly speaking, we have not projected the whole sphere onto the plane, but the sphere minus its north pole (the light source). To correct for this we add a *point at infinity* to the plane—a point approached by the shadows of points on the sphere as they approach the north pole. The point at infinity completes each straight line to a closed curve, so that they too become circles. Thus our second interpretation of spherical geometry models all "lines" by circles, and "angles" by angles between circles. In the next subsection we will see a similar model of non-Euclidean geometry.

Models of Non-Euclidean Geometry

Beltrami (1868) discovered several models of non-Euclidean geometry; that is, of Euclid's basic axioms plus a non-Euclidean parallel axiom stating that *for any line l and a point P outside it, there is more than one line m that does not meet l.* The easiest of Beltrami's models to view in its entirety is the one shown in figure 1.8.

Figure 1.8 : The conformal disk model

In this model, "points" are points in the interior of the disk, "lines" are circular arcs perpendicular to the boundary circle of the disk (counting the straight line segments through the disk center as circles of infinite radius) and "angle" is the angle between circles. As in spherical geometry, triangles are congruent if they have the same angles, so in this picture the disk is filled with infinitely many congruent triangles, each with the angles $\pi/2, \pi/3, \pi/7$. These are the smallest triangles that can tile the non-Euclidean plane and, as in spherical geometry, their area is determined by their angle sum: *π minus the angle sum of a non-Euclidean triangle is* *proportional to its area.*

As with the plane model of spherical geometry, the precise definition of "distance" is complicated. But here one gets a better feel for it because there are so many triangles, each of the same non-Euclidean size. One sees, for example, that infinitely many triangles lie along each "line," so each "line" is of infinite "length." It is even possible to accept that each "line" gives the least "distance" between any two points in the disk, since one counts fewer triangles when travelling on a circular arc perpendicular to the boundary than on any other route. Thus one can understand how the model satisfies the basic axioms of Euclid. But it clearly does *not* satisfy the parallel axiom. If one takes the vertical "line" l through the center of the disk and the point P, say, somewhat to its right, then there are different "lines" m and n through P that do not meet l, as is clear from figure 1.9.

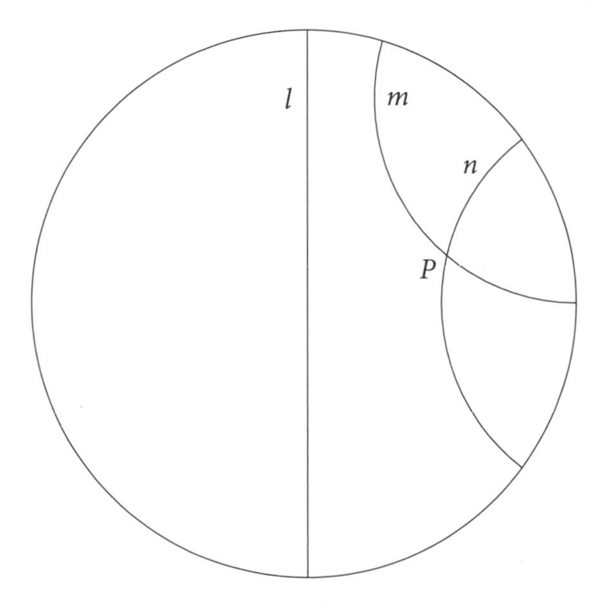

Figure 1.9 : Failure of the parallel axiom

So, when the details of Beltrami's construction are checked, one has *a model for the basic axioms of Euclid plus a counterexample to the parallel axiom.* Therefore, *the parallel axiom does not follow from the other axioms of Euclid*, and hence the theorems equivalent to the parallel axiom (such as the four mentioned in the previous section) likewise do not follow from Euclid's other axioms. However, the equivalences between the parallel axiom and these theorems are provable from Euclid's other axioms. This situation is typical of reverse mathematics: we have a *base theory* which is too weak to prove certain desirable theorems, but strong enough to prove *equivalences* between them.

New Foundations of Geometry and Mathematics

The discovery of non-Euclidean geometry shook the foundations of mathematics, which before the nineteenth century had been implicitly based on Euclid's concepts of "line" and "plane." By creating doubts about the meaning of "line" and "plane," non-Euclidean geometry prompted a search for new foundations in *arithmetic*, since the fundamental properties of numbers were not in doubt.

In particular, the "line" was rebuilt as the system $\mathbb{R}$ of *real numbers*, which has both algebraic and geometric properties. The next few sections describe the emergence of geometry based on, or influenced by, the real number concept. In chapter 2 we will see how the real numbers also became the foundation of analysis.

1.3 VECTOR GEOMETRY

The first major advance in geometry after the Greeks was made by Fermat and Descartes in the 1620s, and published in the *Geometry* of Descartes (1637). Their innovation was to use algebra in geometry, describing lines and curves by equations, thereby reducing many problems of geometry to routine calculations. But before they could "algebraicize" geometry they had to *arithmetize* it, a step that already took them far beyond Euclid. In fact, it was the first step towards a sweeping arithmetization of geometry and analysis that occurred in the nineteenth century.

As every mathematics student now knows, the Euclidean plane is arithmetized by assigning real number *coordinates* x and y to each point P in the plane. The numbers x and y are visualized as the horizontal and vertical distances to P from the origin O, in which case the distance $|OP|$ of P from O is $\sqrt{x^2+y^2}$ by the Pythagorean theorem (figure 1.10). But P can be *defined* as the ordered pair[2] $\langle x, y\rangle$, and its distance from O defined as $\sqrt{x^2+y^2}$. More generally, the distance from $P_1 = \langle x_1, y_1\rangle$ to

[2]In this book I use $\langle a, b\rangle$ to denote the ordered pair of a and b, because (a, b) will be on duty to represent the open interval between a and b.

$P_2 = \langle x_2, y_2 \rangle$ is defined by

$$|P_1P_2| = \sqrt{(x_2 - x_1)^2 + (y_2 - y_1)^2}.$$

Points $\langle x, y \rangle$ lie on a *line* if they satisfy an equation of the form $ax+by+c = 0$ (which is why we call such equations *linear*), and equations for circles are quadratic equations expressing constant distance for a point. For example, the points at distance 1 from O satisfy the equation $x^2 + y^2 = 1$.

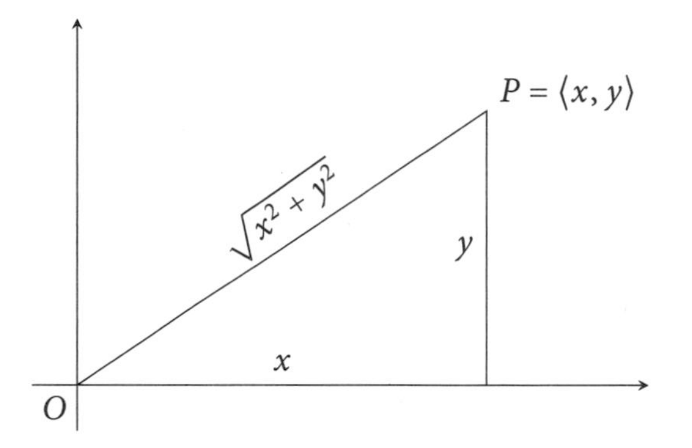

Figure 1.10 : Coordinatizing the plane

Thus one has an easy algebraic translation of all of Euclid's geometry—*and more*, since there is no obstacle, other than algebraic difficulty, to the study of curves satisfying arbitrary polynomial equations. Thus Euclidean geometry and algebraic geometry are not a perfect match. Euclidean geometry ought to be "more linear."

Grassmann's Linear Geometry

The perfect algebraic match for Euclidean geometry was found by Grassmann in the 1840s, in the concept of a *real vector space*. His first works on the subject, Grassmann (1844) and Grassmann (1847) were impenetrable to other mathematicians, and his idea started to gain traction only when Peano (1888) gave axioms for real vector spaces.

Definition. A real vector space is a set V of objects called *vectors* (denoted by boldface letters), which includes a vector $\mathbf{0}$ called the *zero vector*, and for each $\boldsymbol{u} \in V$ a vector $-\boldsymbol{u}$ called the *negative* of $\boldsymbol{u}$. V has operations of *addition* and *scalar multiplication* (by $a, b, c, \ldots \in \mathbb{R}$) satisfying the following conditions:

$$
\begin{aligned}
\boldsymbol{u}+\boldsymbol{v} &= \boldsymbol{v}+\boldsymbol{u}\\
\boldsymbol{u}+(\boldsymbol{v}+\boldsymbol{w}) &= (\boldsymbol{u}+\boldsymbol{v})+\boldsymbol{w}\\
\boldsymbol{u}+\boldsymbol{0} &= \boldsymbol{u}\\
\boldsymbol{u}+(-\boldsymbol{u}) &= \boldsymbol{0}\\
1\boldsymbol{u} &= \boldsymbol{u}\\
a(\boldsymbol{u}+\boldsymbol{v}) &= a\boldsymbol{u}+a\boldsymbol{v}\\
(a+b)\boldsymbol{u} &= a\boldsymbol{u}+b\boldsymbol{u}\\
a(b\boldsymbol{u}) &= (ab)\boldsymbol{u}
\end{aligned}
$$

Typically $V = \mathbb{R}^n = \{\langle x_1, \ldots, x_n\rangle : x_1, \ldots, x_n\}$, with $\boldsymbol{0}$ the origin, + the usual sum of n-tuples, and scalar multiplication by $a \in \mathbb{R}$ given by

$$a\langle x_1, \ldots, x_n\rangle = \langle ax_1, \ldots, ax_n\rangle.$$

This vector space is called the *real n-dimensional affine space*. It is not yet a Euclidean space because it has no concept of distance or angle, but it has considerable geometric content. $\mathbb{R}^n$ has lines, including parallel lines, and also a concept of "length in a given direction." For example, one can say that $\frac{1}{2}\boldsymbol{v} \in \mathbb{R}$ is the *midpoint* of the line from $\boldsymbol{0}$ to $\boldsymbol{v}$, and in general $a\boldsymbol{v}$ is a times as far from $\boldsymbol{0}$ as $\boldsymbol{v}$ is. Another concept that makes sense in vector geometry is that of *center of mass*. In particular, the center of mass of the triangle with vertices $\boldsymbol{u}, \boldsymbol{v}, \boldsymbol{w}$ is the point $\frac{1}{3}(\boldsymbol{u}+\boldsymbol{v}+\boldsymbol{w})$.

To promote vector geometry to Euclidean geometry one adds the concept of *inner product* of vectors $\boldsymbol{u}$ and $\boldsymbol{v}$, written $\boldsymbol{u}\cdot\boldsymbol{v}$:

Definition. If $\boldsymbol{u} = \langle u_1, \ldots, u_n\rangle$ and $\boldsymbol{v} = \langle v_1, \ldots, v_n\rangle$ then

$$\boldsymbol{u}\cdot\boldsymbol{v} = u_1v_1 + \cdots + u_nv_n.$$

In particular, in $\mathbb{R}^2$ we have

$$\boldsymbol{u}\cdot\boldsymbol{u} = u_1^2 + u_2^2,$$

so the Euclidean length $|\boldsymbol{u}|$ of $\boldsymbol{u}$ is given by $|\boldsymbol{u}| = \sqrt{\boldsymbol{u}\cdot\boldsymbol{u}}$. As Grassmann (1847) remarked, the definition of inner product makes the Pythagorean theorem true almost by definition.

The Euclidean angle concept also derives from the inner product because

$$\boldsymbol{u}\cdot\boldsymbol{v} = |\boldsymbol{u}||\boldsymbol{v}|\cos\theta,$$

where θ is the angle between the lines from $\mathbf{0}$ to $\boldsymbol{u}$ and $\boldsymbol{v}$ respectively. Thus Grassmann (1847) found another way to describe Euclidean geometry as a "base theory" plus the "right axiom" to derive the Pythagorean theorem. Interestingly, his base theory (the vector space axioms) admits extension by a different axiom that gives *non*-Euclidean geometry.

Making a Vector Space Non-Euclidean

The key property of Grassmann's inner product is that it is *positive definite*; that is, $|\boldsymbol{u}|^2 = \boldsymbol{u} \cdot \boldsymbol{u} > 0$ if $\boldsymbol{u} \neq \mathbf{0}$, so every nonzero vector has positive length. Einstein's theory of special relativity motivated Minkowski (1908) to introduce a *non*-positive definite inner product on the space $\mathbb{R}^4$ of spacetime vectors $\langle t, x, y, z\rangle$, namely

$$\langle t_1, x_1, y_1, z_1\rangle \cdot \langle t_2, x_2, y_2, z_2\rangle = -t_1t_2 + x_1x_2 + y_1y_2 + z_1z_2.$$

With the Minkowski inner product $\boldsymbol{u} = \langle t, x, y, z\rangle$ has "length" $|\boldsymbol{u}|$ given by

$$|\boldsymbol{u}|^2 = -t^2 + x^2 + y^2 + z^2,$$

which clearly is zero or negative for many vectors. To make visualization easier we consider the corresponding concept of length on the space $\mathbb{R}^3$ of vectors $\boldsymbol{u} = \langle t, x, y\rangle$, namely

$$|\boldsymbol{u}|^2 = -t^2 + x^2 + y^2.$$

This means that in $\mathbb{R}^3$ we have a "sphere[3] of radius $\sqrt{-1}$ about O," consisting of the vectors $\boldsymbol{u} = \langle t, x, y\rangle$ such that

$$-t^2 + x^2 + y^2 = -1.$$

This surface in $\mathbb{R}^3$ is the *hyperboloid* $x^2 + y^2 - t^2 = 1$.

It turns out that the Minkowski distance on the surface of the hyperboloid gives a non-Euclidean geometry—the same as that of the Beltrami model in the previous section. Figure 1.11, which is derived from a picture by Konrad Polthier of the Freel University of Berlin, shows the connection between the two. The tiling of the disk projects to a tiling of the hyperboloid by triangles that are congruent in the sense of Minkowski distance.

[3]In a remarkable prophecy, Lambert (1766) conjectured that there might be a geometry on the sphere of imaginary radius for which the angle sum of a triangle is less than π, and where the area of a triangle is proportional to π minus its angle sum. This is indeed what happens in Beltrami's non-Euclidean geometry.

Figure 1.11 : The hyperboloid model of non-Euclidean geometry

1.4 HILBERT'S AXIOMS

Euclid's *Elements* is the first organized presentation of mathematics that survives from ancient times. It is best known for its treatment of geometry, deducing theorems from axioms in a style that became standard for mathematics until the nineteenth century. Then the discovery of non-Euclidean geometry put Euclid's geometry under the microscope, and by the late nineteenth century his axioms were found to have some gaps. But this only strengthened the movement towards axiomatization. The gaps in Euclid were filled by Hilbert (1899) and, in the meantime, axiomatic treatments of number theory and algebra were given by Dedekind, Peano, and others.

Euclid also gave a deductive treatment of numbers in the *Elements*, but it was complicated by the Greek discovery of irrationality, which was thought to disqualify some geometric quantities (such as the diagonal $\sqrt{2}$ of the unit square) from being numbers at all. Irrational quantities were not fully reconciled with whole or rational numbers until the publication of the Dedekind (1872) book on irrational numbers. Dedekind found that Euclid had been on the right track—the only new idea needed to make his theory of irrational quantities part of his theory of numbers was acceptance of *infinite sets* of rational numbers (see section 1.5).

The two main threads of the *Elements*, geometry and the real numbers, were combined in the *Grundlagen der Geometrie* (foundations of geometry) of Hilbert (1899). Here, Hilbert not only filled the gaps in Euclid's geometric axioms, he also introduced two axioms that complete a geometric path to the real number system $\mathbb{R}$. This was a historic achievement, though Hilbert's path is not the best for all mathematical purposes. The *arithmetization* path to real numbers via the rational numbers ultimately proved more useful for analysis, and we will take it up again in chapter 2.

Hilbert (1899) found that Euclid's geometry and the arithmetic of real numbers follow from 17 axioms, described below. All but two of them are purely geometric. The exceptions are the *Archimedean axiom*, which says no line segment is "infinitely large" compared with another, and the *completeness axiom*, which says there are no "gaps" in the points on a line. (These two axioms were not needed by Euclid, who considered only points constructible by ruler and compass.) Their purpose is to prove that any line satisfying the axioms is essentially the line $\mathbb{R}$ of real numbers. It follows that any plane satisfying the axioms is essentially the plane of Descartes, so Euclid's geometry has really only one model—the plane of pairs of real numbers.

This very satisfying convergence of the geometric and arithmetic viewpoints comes about because Hilbert's geometric axioms yield not just Euclid's geometric theorems—they also yield *algebra*, which Euclid did not foresee. In fact, algebraic structure arises in stages corresponding to axiom *groups*, which Hilbert introduces one by one.

Axioms of incidence. These relate lines and points. They include Euclid's axiom that two points determine a line, and a form of the parallel axiom: for any line l and point $P \notin l$ there is exactly one line m through P not meeting l. Also (which went without saying in Euclid) each line has at least two points, and there are three points not in a line.

Axioms of order. The first three of these axioms say the obvious things about the order of three points on a line: if B is between A and C then it is also between C and A; any A and C have a point B between them; for any three points, one is between the other two. The fourth, called *Pasch's axiom*, is about the plane: a line meeting one side of a triangle at an internal point meets exactly one of the other sides.

Axioms of congruence. The first five of these axioms are about equality of line segments or angles, and the addition of line segments. They state the existence and uniqueness of line segments or angles equal to given ones, at a given position. They also say (as Euclid put it) "things equal to the same thing are equal to each other." The last congruence axiom is the SAS criterion for congruence of triangles.

Circle intersection axiom. Two circles meet if one of them contains points both inside and outside the other. (Euclid overlooked this axiom, even though he assumed it in his very first proposition, constructing an equilateral triangle.) Note that the points "inside" a circle of radius r are those at distance $< r$ from its center.

Archimedean axiom. For any nonzero line segments AB and CD there is a natural number n such that n copies of AB are together greater than CD.

Completeness axiom. Suppose the points of a line l are divided into two nonempty subsets $\mathcal{A}$ and $\mathcal{B}$ such that no point of $\mathcal{A}$ is between two points of $\mathcal{B}$ and no point of $\mathcal{B}$ is between two points of $\mathcal{A}$. Then there is a unique point P, in either $\mathcal{A}$ or $\mathcal{B}$, that lies between any other two points, of which one is in $\mathcal{A}$ and the other is in $\mathcal{B}$. (Thus, there is no "gap" between $\mathcal{A}$ and $\mathcal{B}$.)

These axioms give precise meaning to the idea of a theorem being *equivalent* to the parallel axiom: namely, the equivalence is provable in the *base theory* of Hilbert's axioms *minus* the parallel axiom. All theorems previously thought to be equivalent to the parallel axiom (such as those mentioned in section 1.1) are equivalent to it in this sense. As suggested at the end of section 1.2, proving equivalences in a weaker system is the hallmark of *reverse mathematics*. We will see further historical examples in the later sections of this chapter. Today, the idea has been most fully developed in systems of analysis, and we will see some its main results in chapters 6 and 7.

Algebraic Content of Hilbert's Axioms

The incidence axioms allow us to define sum and product of points on a line by means of the constructions shown in figures 1.12 and 1.13.

The sum construction chooses a point 0 on the line then, for any points a and b on the line, constructs a point $a + b$ with the help of the parallels shown. In effect, the parallels allow the point b to be "translated" along the line by the distance between 0 and a.

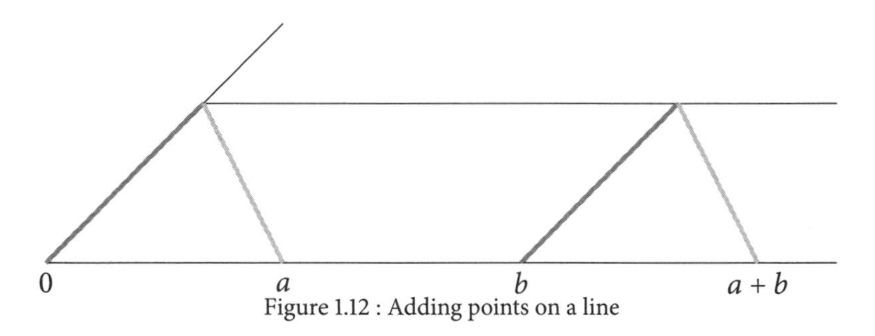

Figure 1.12 : Adding points on a line

The product construction also requires a point 1 on the line (the "unit of length"), and various parallels now allow us to "magnify" the distance from 0 to b by the distance from 0 to a, producing the point ab.

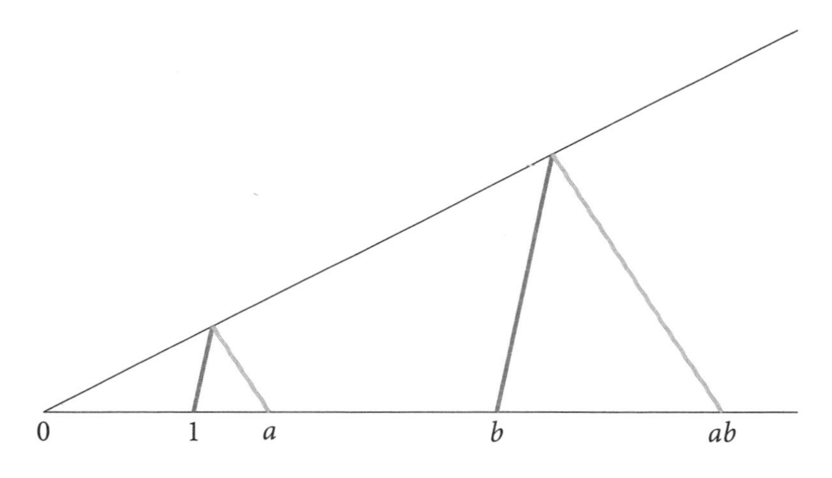

Figure 1.13 : Multiplying points on a line

With the help of the congruence axioms one can prove that the sum and product operations just defined have the following algebraic properties, the *field properties* (also used as the *axioms* that define a field):

$$a+b=b+a \qquad a\cdot b=b\cdot a \qquad \text{(commutativity)}$$

$$a+(b+c)=(a+b)+c \qquad a\cdot(b\cdot c)=(a\cdot b)\cdot c \qquad \text{(associativity)}$$

$$a+0=a \qquad a\cdot 1=a \qquad \text{(identity)}$$

$$a+(-a)=0 \qquad a\cdot a^{-1}=1 \text{ for } a\neq 0 \qquad \text{(inverse)}$$

$$a\cdot(b+c)=a\cdot b+a\cdot c \qquad \text{(distributivity)}$$

It is easiest to deduce the field properties from the congruence axioms, but there is in fact a pure incidence axiom—the so-called *Pappus theorem*—from which all the field properties follow with the help of the other incidence axioms.[4] Thus the algebraic structure of a field emerges from axioms that Euclid almost completely overlooked: the incidence axioms describing how points and lines interact.

The order axioms give the points on a line an ordering, $\leq$, with the properties that, for any a, b, c:

- $a \leq a$,
- if $a \neq b$ then either $a < b$ or $b < a$, but not both,
- if $a \leq b$ and $b \leq c$ then $a \leq c$.

The order relation meshes with the field properties to produce an *ordered field*. Its defining properties, beyond the field properties above, are that:

- if $a \leq b$ then $a + c \leq b + c$,
- if $0 \leq a$ and $0 \leq b$ then $0 \leq ab$.

Finally, the Archimedean and completeness axioms say that the order relation is *Archimedean* and *complete* in the sense described by those axioms. It can be proved that a *complete Archimedean ordered field is isomorphic to the field* $\mathbb{R}$ *of real numbers*. Given such a field $\mathbb{F}$, the idea of the proof is to build a copy of $\mathbb{R}$ inside $\mathbb{F}$ in the following stages. (Readers not yet familiar with the construction of the real numbers as Dedekind cuts may wish to take these steps on faith and confirm them later when reading chapter 2.)

[4]Incidentally, the field properties can be proved in the setting of *projective geometry*, where all axioms are incidence axioms and the parallel axiom is replaced by the axiom that any two lines meet in a unique point. The above constructions can be carried out in this setting when we call one line a "line at infinity" and call lines "parallel" when they meet on the line at infinity.

1. From the element $1 \in \mathbb{F}$ build the "positive integers" of $\mathbb{F}$, namely

$$1, \quad 1+1, \quad 1+1+1, \quad 1+1+1+1, \ldots,$$

using the + operation of $\mathbb{F}$.
2. Build "integers" of $\mathbb{F}$ using 0 and the − operation.
3. Build "rational numbers" of $\mathbb{F}$ using inverse and product operations.
4. Use the order and completeness properties of $\mathbb{F}$ to build "real numbers" of $\mathbb{F}$ as Dedekind cuts in the "rational numbers" of $\mathbb{F}$.
5. Check that the "real numbers" of $\mathbb{F}$ exhaust the members of $\mathbb{F}$ and have the same properties as the actual real numbers.

This proof shows that any complete Archimedean ordered field is essentially the "same" as $\mathbb{R}$, so every line in Hilbert's geometry is essentially the real number line. The next question is: how well do we understand $\mathbb{R}$?

1.5 WELL-ORDERING AND THE AXIOM OF CHOICE

In Book V of the *Elements*, Euclid gave a very sophisticated treatment of the geometric line and its relationship to the rational numbers. He stopped short of declaring irrational points to *be* numbers, but he essentially showed that each point is approximated arbitrarily closely by rational numbers. This means that each point is *determined by* rational numbers (those to the left of the point, for example), so *we need only accept infinite sets as mathematical objects* in order to view points as arithmetical objects.

However, until the mid-nineteenth century, most mathematicians rejected the idea of infinite sets as mathematical objects. They were influenced by the ancient Greek distinction between "potential" and "actual" infinity. For example, it was permissible to view the natural numbers as an open-ended process—start with 0 and keep adding 1—but not as a completed or "actual" entity $\mathbb{N} = \{0, 1, 2, \ldots\}$. Today, this seems a rather hair-splitting distinction, because—as far as anyone knew in the mid-nineteenth century—all infinite sets could be viewed as "potential" infinities.

For example, the integers, $\mathbb{Z}$, can be viewed as a potential infinity by listing them in the order

$$0, \quad 1, \quad -1, \quad 2, \quad -2, \quad 3, \quad -3, \quad \ldots.$$

The positive rationals can likewise be viewed as a potential infinity by listing them in the order

$$\frac{1}{1}, \quad \frac{2}{1}, \quad \frac{1}{2}, \quad \frac{3}{1}, \quad \frac{1}{3}, \quad \frac{4}{1}, \quad \frac{3}{2}, \quad \frac{2}{3}, \quad \frac{1}{4}, \quad \ldots .$$

(The rule is to list fractions m/n in order of the sums $m + n$: first those with $m+n = 2$, then those with $m+n = 3$, then those with $m+n = 4$, and so on.) And then we can view *all* the rationals, $\mathbb{Q}$, as a potential infinity by listing positive and negative elements alternately as we did with $\mathbb{Z}$:

$$0, \quad \frac{1}{1}, \quad -\frac{1}{1}, \quad \frac{2}{1}, \quad -\frac{2}{1}, \quad \frac{1}{2}, \quad -\frac{1}{2}, \quad \frac{3}{1}, \quad -\frac{3}{1}, \quad \frac{1}{3}, \quad -\frac{1}{3}, \quad \ldots .$$

Thus $\mathbb{N}$, $\mathbb{Z}$, and $\mathbb{Q}$, which we now regard as sets, could all be finessed as "potential" infinities by mathematicians who were fastidious about the distinction between potential and actual.

A much more serious problem arose in 1874, when Cantor showed that $\mathbb{R}$ is not by any means a potential infinity.

Uncountability

The means by which we showed the sets $\mathbb{N}$, $\mathbb{Z}$, and $\mathbb{Q}$ to be potential infinities was by counting, or *ordering* their members in a sequence:

1st member, 2nd member, 3rd member, ...

—with an implied process for counting members that reaches each member at some finite stage. Cantor (1874) showed that $\mathbb{R}$ is *uncountable* in the sense that *no such ordering of* $\mathbb{R}$ *exists.*

He showed that any sequence $x_1, x_2, x_3, \ldots$ of real numbers fails to include some real number x. In fact, given the decimal expansions of $x_1, x_2, x_3, \ldots$ we can *compute* the decimal expansion of x. For example, we can use the rule:

$$n\text{th decimal digit of } x = \begin{cases} 1 & \text{if } n\text{th decimal digit of } x_n \neq 1 \\ 2 & \text{if } n\text{th decimal digit of } x_n = 1. \end{cases}$$

Then $x \neq$ each x_n because x differs from x_n in the nth decimal place.

Thus if we accept $\mathbb{R}$ we have to accept it as an *actual* infinity. The proof given here is essentially one given by Cantor (1891). It is, incidentally, a harbinger of many proofs about $\mathbb{R}$ that we will see later in this book.

Given an arbitrary object, such as a sequence or a function, we prove existence of some other object by *computing it from* the given object. The computation of one object relative to others is seldom noticed in classical analysis—in fact, many mathematicians have thought that Cantor's proof is *nonconstructive*—but it is important, as we will see in later chapters.

Well-ordering

Cantor's theorem shows that $\mathbb{R}$ cannot be ordered in the simple way that $\mathbb{N}$, $\mathbb{Z}$, and $\mathbb{Q}$ can: 1st member, 2nd member, 3rd member, ... Nevertheless Cantor (1883) stated his belief in a more general kind of order:

> In a later article I shall discuss the law of thought that says that it is always possible to bring any *well-defined* set into the *form* of a *well-ordered* set—a law which seems to me fundamental and momentous and quite astonishing. (Ewald (1996), vol. II, p. 886)

Cantor called a set S well-ordered if the ordering is such that every nonempty subset T of S has a *least* member. This is clearly the case for the orderings of $\mathbb{N}$, $\mathbb{Z}$, and $\mathbb{Q}$ above, where each member is labeled with a positive integer (take the member of T whose integer is least). It is also the case for the following ordering of $\mathbb{Z}$,

$$0, \quad 1, \quad 2, \quad 3, \quad \ldots, \quad -1, \quad -2, \quad -3, \quad -4, \quad \ldots,$$

in which 0 and the positive integers precede all the negative integers. If T is a nonempty subset of $\mathbb{Z}$ then the least member of T in the above ordering is

the least non-negative integer in T, if T has any non-negative members,

or

the greatest negative integer in T, if T has only negative members.

When it comes to $\mathbb{R}$, however, all human ingenuity fails to find a well-ordering of the real numbers. The usual ordering $<$ fails dismally, because subsets such as $\{x \in \mathbb{R} : 0 < x\}$ have no least member. Thus Cantor was very bold to assume that well-orderings exist for all "well-defined" sets—which surely include $\mathbb{R}$.

The Well-ordering Theorem and Zermelo's Axioms

Cantor perhaps thought that his "fundamental law of thought" should be an axiom of set theory. But he did not suggest a set of axioms for set theory, so it remained unclear whether well-ordering should be an axiom or a theorem. The picture became clearer when Zermelo (1904) *proved* well-ordering from an intuitively simpler assumption, now known as the *axiom of choice.*

Axiom of choice (AC). For any set X of nonempty sets x there is a *choice function*; that is, a function f such that $f(x) \in x$ for each $x \in X$.

To provide a precise framework for his proof (and at the same time to clear up some doubts about the foundations of set theory) Zermelo (1908) gave the first set of axioms for set theory. Within his system, now called Z, it was possible to prove that AC is *equivalent* to the well-ordering theorem. Fraenkel (1922) strengthened one of Zermelo's axioms, obtaining a system now known as ZF set theory.

The ZF axioms of set theory have remained stable since 1922 and have become generally accepted as a foundation for all of mainstream mathematics, at least when supplemented by AC. Indeed, it was proved in ZF that AC is equivalent to many sought-after theorems, including the well-ordering theorem, that were apparently not provable outright in ZF.

This put AC in a position, relative to ZF, like that of the parallel axiom relative to the other axioms of Euclid (or, more precisely, relative to the other axioms of Hilbert). Theorems proved equivalent to AC in ZF were not of clear interest until it was known that AC is *not* provable in ZF. This was done by Cohen (1963). As Beltrami in 1868 did for the parallel axiom, Cohen showed the unprovability of AC by constructing a *model* of ZF in which AC is false. His construction, like Beltrami's, was a breakthrough that completely changed the face of the subject. It is unfortunately too technical to be described in this book, but we can describe some of its consequences.

A Mathematical Equivalent of the Axiom of Choice

Like the parallel axiom in geometry, AC in set theory occupies an important position "above" the basic (ZF) axioms. ZF cannot prove AC, but ZF is a good base theory because it can prove that AC is equivalent to many other interesting statements of set theory and general mathematics. In this sense, AC is the "right axiom" to prove these statements. As we know, one such statement is the well-ordering theorem. Another is the following property of *vector spaces* over an arbitrary field $\mathbb{F}$. (We defined a *real* vector space in section 1.3. The definition of an arbitrary vector space is the same, except with $\mathbb{F}$ in place of $\mathbb{R}$.)

Existence of a vector space basis. *Any vector space V has a* basis; *that is, a subset U of vectors $\boldsymbol{u}$ such that:*

(i) For any $\boldsymbol{v} \in V$ there are $\boldsymbol{u}_1, \ldots, \boldsymbol{u}_k \in U$ and $f_1, \ldots, f_k \in \mathbb{F}$ such that $\boldsymbol{v} = f_1\boldsymbol{u}_1 + \cdots + f_k\boldsymbol{u}_k$. ("$U$ spans V.")

(ii) For any distinct $\boldsymbol{u}_1, \ldots, \boldsymbol{u}_k \in U$ and $f_1, \ldots, f_k \in \mathbb{F}$, $\mathbf{0} = f_1\boldsymbol{u}_1 + \cdots + f_k\boldsymbol{u}_k$ if and only if $f_1 = \cdots = f_k = 0$. ("U is an independent set.")

The existence of a basis is clear for finite-dimensional real vector spaces, where we can take the basis vectors $\boldsymbol{u}$ to be the unit points on the coordinate axes. The first case in which a basis is hard to find—in fact utterly mysterious—is when $\mathbb{R}$ is viewed as a vector space over $\mathbb{Q}$. Hamel (1905) showed existence of a basis with the help of a well-ordering of $\mathbb{R}$, but the so-called *Hamel basis* is no easier to define than a well-ordering of $\mathbb{R}$ itself.

Thus it is no surprise that all proofs of the existence of a basis for an arbitrary vector space depend on AC. We now know that AC is unavoidable because Blass (1984) showed that existence of such a basis can be proved *equivalent* to AC in ZF.

1.6 LOGIC AND COMPUTABILITY

The previous sections of this chapter suggest that the real number system $\mathbb{R}$ is an essential part of the foundations of mathematics. When we turn to analysis, in the next chapter, the unavoidability of $\mathbb{R}$ will become even more obvious. At the same time, we have seen that our understanding of $\mathbb{R}$ can never be complete, if only because of the uncountability of $\mathbb{R}$.

Since we cannot list all real numbers we certainly cannot list all *facts* about real numbers, let alone set up an axiom system for proving them. This observation is the first step on the road towards the profound theorems of Gödel (1931) and Turing (1936) about unprovable theorems and unsolvable algorithmic problems—a road we will describe in more detail in chapter 4.

Gödel's theorem rules out any possibility of a complete axiom system for analysis. Yet it also presents an opportunity. If we are lucky we may be able to find a *base theory* for analysis, in which we can prove that sought-after theorems are equivalent to certain axioms—axioms that play a role, like that of the parallel axiom in geometry or AC in set theory, of attracting desirable theorems into their "orbit" of equivalent theorems.

This indeed is what happens. We now know a good base theory, called RCA_0, and at least four *set existence axioms* that play this role for theorems of analysis. Moreover, the axioms are of increasing strength, in the sense that each implies the one before, so they classify theorems of analysis by increasing strength. The crucial axioms state "set existence" rather than "real number existence" because it is technically convenient to encode real numbers by sets of natural numbers (see next chapter for details). The axioms in question state there is a set of natural numbers n corresponding to each property $\varphi(n)$ in a certain class.

For RCA_0 we assert set existence for the class of *computable* properties $\varphi(n)$. These are the properties for which there is an algorithm that decides, for each n, whether $\varphi(n)$ holds. It turns out, because *non*computable properties exist, that RCA_0 is too weak to prove many important theorems of analysis. But RCA_0 can prove many equivalences, since these often involve computing an object (such as a sequence or a function) from a given object. For example, RCA_0 cannot prove the Bolzano-Weierstrass theorem, but it can prove that Bolzano-Weierstrass is equivalent to an axiom stating the existence of sets realizing each *arithmetically definable* property $\varphi(n)$. Thus, if we add the latter axiom to RCA_0, we obtain a stronger system in which Bolzano-Weierstrass is provable.

In this way we find, rather surprisingly, that most of the well-known theorems of analysis can be assigned a precise level of "strength." They are either at the lowest level—provable in RCA_0—or at a higher level represented by one of four set existence axioms. In this book we focus mainly on the lower three levels, where most of the well-known theorems of analysis are known to reside (see chapters 6 and 7).

Arithmetization

From the discussion above we can see that a study of arithmetic and computation will be needed before we can define the system RCA_0. Arithmetic itself is axiomatized in a fairly standard way that goes back to the *Peano axioms* of Peano (1889). But before that we have to talk about *arithmetization*—both in the nineteenth-century sense of making analysis "arithmetical," and in the 1930s sense of making logic and computation "arithmetical."

The remarkable convergence of analysis and computation to a common source in arithmetic is what makes the reverse mathematics of analysis possible. The arithmetization of analysis is discussed in chapter 2, computation is discussed in chapter 4, and its arithmetization in chapter 5. We also give a refresher course on the real numbers and continuity in chapter 3, including classical proofs of the best-known theorems.

CHAPTER 2

Classical Arithmetization

At the International Congress of Mathematicians in Paris in 1900, Henri Poincaré summed up the situation in the foundations of analysis as follows:

> Today in analysis there remain only natural numbers or finite or infinite systems of natural numbers ... Mathematics, as one says, has been arithmetized. (Poincaré (1902), p. 120)

Poincaré was speaking at the end of a century of progress and upheaval in mathematics. As we know from chapter 1, the nineteenth century brought to light new geometries, new algebras, new concepts of number and function, and new infinities—for which the traditional foundation of mathematics, in Euclid, was clearly inadequate.

In the hope of building a secure foundation for the vast new edifice of mathematics, mathematicians turned to the system $\mathbb{N}$ of natural numbers $0, 1, 2, 3, \ldots$. From them (and from *sets* of them) they rebuilt the real and complex numbers, functions, and the geometric objects that were then the main concern of mathematicians. This was the project of *arithmetization*.

In this chapter we describe how one proceeds from natural to rational numbers, then to real and complex numbers, and to continuous functions—thus arithmetizing the foundations of analysis and geometry. Then we turn to the foundations of the natural numbers themselves, the *Peano axioms*, which gives a first glimpse of the logic underlying the arithmetization project.

2.1 FROM NATURAL TO RATIONAL NUMBERS

Integers

We assume for the moment that the natural numbers $0, 1, 2, 3, \ldots$ and their operations of plus $(+)$ and times $(\cdot)$ are given (section 2.6 explains further). From them we create the system $\mathbb{Z}$ of *integers* as ordered pairs $\langle m, n\rangle$ of natural numbers m, n, where the intended interpretation of $\langle m, n\rangle$ is $m - n$. This means that the negative integer -1 is represented by infinitely many pairs:

$$-1 = \langle 0, 1\rangle = \langle 1, 2\rangle = \langle 2, 3\rangle = \langle 3, 4\rangle = \cdots,$$

however, we can tell when two pairs represent the same integer by a test involving only natural numbers and addition, namely:

$$\langle m_1, n_1\rangle = \langle m_2, n_2\rangle \Leftrightarrow m_1 + n_2 = m_2 + n_1.$$

The creation of negative numbers from pairs of natural numbers is not simply a pure mathematical abstraction. The same idea has been used for centuries in *double entry bookkeeping*, where an account balance (which may be negative) is expressed by a pair of positive numbers, the amounts of credit and debit. This idea goes back to Pacioli (1494), and its role in the history of negative numbers is described in Ellerman (2014).

It is also easy, with the intended interpretation in mind, to extend the $+$ and $\cdot$ functions from natural numbers to the integers. Namely,

$$\langle m_1, n_1\rangle \cdot \langle m_2, n_2\rangle = \langle m_1m_2 + n_1n_2, m_1n_2 + m_2n_1\rangle.$$

Notice that this definition answers the vexed question of what $(-1) \cdot (-1)$ is, because -1 is represented by the pair $\langle 0, 1\rangle$ and

$$\langle 0, 1\rangle \cdot \langle 0, 1\rangle = \langle 0 \cdot 0 + 1 \cdot 1, 0 \cdot 1 + 0 \cdot 1\rangle = \langle 1, 0\rangle,$$

which represents the number 1.

And of course (because this is the purpose of negative integers) we now have the *subtraction* operation defined for all integers:

$$\langle m_1, n_1\rangle - \langle m_2, n_2\rangle = \langle m_1 + n_2, m_2 + n_1\rangle.$$

There are a couple of tedious, but straightforward, things to check.

1. That +, −, and · are *well defined*; that is, independent of the choice of representatives.
2. That +, −, and · have the desired algebraic properties—the so-called *ring axioms*—which are the following (where $-a$ stands for $0 - a$).

$$a + b = b + a \qquad a \cdot b = b \cdot a \quad \text{(commutativity)}$$

$$a + (b + c) = (a + b) + c \qquad a \cdot (b \cdot c) = (a \cdot b) \cdot c \quad \text{(associativity)}$$

$$a + 0 = a \qquad a \cdot 1 = a \quad \text{(identity)}$$

$$a + (-a) = 0 \quad \text{(inverse)}$$

$$a \cdot (b + c) = a \cdot b + a \cdot c \quad \text{(distributivity)}$$

These depend on related properties of the natural numbers, discussed further in section 2.6.

Rational Numbers

From the integers we create the system $\mathbb{Q}$ of rational numbers as ordered pairs $\langle i, j \rangle$ of integers, where $j \neq 0$. The intended interpretation of $\langle i, j \rangle$ is i/j, so again there are infinitely many representations of each rational number—but of course we are used to the fact that each rational number is expressed by infinitely many fractions, differing only by a common factor in i and j.

We are also used to the rule for multiplying fractions,

$$\langle i_1, j_1 \rangle \cdot \langle i_2, j_2 \rangle = \langle i_1 i_2, j_1 j_2 \rangle,$$

and the rule for adding them by taking the "common denominator" $j_1 j_2$:

$$\langle i_1, j_1 \rangle + \langle i_2, j_2 \rangle = \langle i_1 j_2 + i_2 j_1, j_1 j_2 \rangle.$$

The latter rule—a notorious stumbling block in elementary mathematics—can be better appreciated when one tests the claim that + is well-defined for rational numbers. Since

$$\langle i_1, j_1 \rangle = \langle m i_1, m j_1 \rangle \quad \text{for any integer } m \neq 0,$$

$$\langle i_2, j_2 \rangle = \langle n i_2, n j_2 \rangle \quad \text{for any integer } n \neq 0,$$

their sum is

$$\langle m i_1 n j_2 + n i_2 m j_1, m j_1 n j_2 \rangle.$$

The latter expression is indeed independent of m and n, because of the common factor mn in both members of the pair.

Algebraic Properties

The definitions of integers and rational numbers above show why questions about them can, in principle, be reduced to questions about natural numbers and their addition and multiplication. This is what it means to say that the natural numbers are a *foundation* for the integer and rational numbers.

In practice, of course, we prefer to work with integer or rational numbers. They admit not just addition and multiplication, but also subtraction and division, and their rules for calculation are simpler. We could say that the rational numbers have better algebraic properties than the natural numbers. These are the *field* properties—already mentioned in the geometric context of section 1.4—which are the above ring properties plus the *multiplicative inverse* property: $a \cdot a^{-1} = 1$ when $a \neq 0$. Nevertheless, the rules for calculating with rational numbers can be traced back to the natural numbers, as we will see in section 2.6.

The next steps in the arithmetization project go beyond algebra. By admitting *sets* of rational numbers we can enlarge the number system to one that admits certain *infinite* operations, such as forming infinite sums. This is crucial to building a foundation for analysis.

2.2 FROM RATIONALS TO REALS

Since the time of Pythagoras it has been known that rational points do not fill the line. They do not include the point $\sqrt{2}$, for example. To fill gaps in the line, mathematicians have devised various infinite processes that create new points, such as infinite sums, infinite decimals, and infinite continued fractions. But the most direct way to fill the gaps in the rational numbers is one that stems from an idea of Dedekind (1872): *fill each gap by the gap itself!*

We can view each gap (or *cut* as Dedekind called it) in the set $\mathbb{Q}$ of rational numbers as a division of $\mathbb{Q}$ into two disjoint subsets, L and U ("lower" and "upper"), where L has no greatest member, U has no least member, and each member of L is less than all members of U. Thus gaps, too, are infinite objects, but they are the objects most suitable for completing the number line. There is $\mathbb{Q}$, and there are gaps in $\mathbb{Q}$, and together they fill the line.

This may seem at first like a trick, a play on words, but the trick works extremely well. It not only completes the set $\mathbb{Q}$ of rational numbers to the gapless set $\mathbb{R}$ of real numbers, but also ensures that $\mathbb{R}$ is a field, because the field properties are directly inherited from $\mathbb{Q}$. This becomes clear if we extend the concept of Dedekind cut to include rational numbers as well, by dropping the condition that L have no greatest member. Indeed, each real number can be represented by a lower set L alone, which we will call a *lower Dedekind cut*:

Definition. A real number is a set L of rational numbers that is bounded above and "closed downward": that is, if $s \in L$ and rational $t < s$ then $t \in L$.

We illustrate how $\mathbb{R}$ inherits the field properties from $\mathbb{Q}$ in the case of addition.

Definition. If L_1 and L_2 are real numbers, their *sum* $L_1 + L_2$ is defined by

$$L_1 + L_2 = \{s_1 + s_2 : s_1 \in L_1 \text{ and } s_2 \in L_2\}.$$

From this definition the commutative, associative, identity, and inverse laws for addition follow immediately. For example, here is why associativity holds:

$$\begin{aligned} L_1 + (L_2 + L_3) &= \{s_1 + (s_2 + s_3) : s_1 \in L_1 \text{ and } s_2 \in L_2 \text{ and } s_3 \in L_3\} \\ &\qquad\qquad\qquad\qquad\qquad\qquad \text{(by definition of sum)} \\ &= \{(s_1 + s_2) + s_3 : s_1 \in L_1 \text{ and } s_2 \in L_2 \text{ and } s_3 \in L_3\} \\ &\qquad\qquad\qquad\qquad\qquad\qquad \text{(by associativity in } \mathbb{Q}) \\ &= (L_1 + L_2) + L_3 \qquad\qquad\qquad \text{(by definition of sum)} \end{aligned}$$

The definition of product is a little more tricky, since multiplying numbers in the negative part of a cut will produce arbitrarily large positive rationals, so the resulting set is not a lower Dedekind cut. One way around this problem is to define *positive* reals via cuts in the positive rationals, and to define their products first. Then treat arbitrary reals as pairs of positive reals as we did when constructing integers from the rational numbers. It remains true that the field properties of multiplication for reals follow immediately from those for rationals.

Thus the real numbers behave as we expect "numbers" to behave. Now let us see how well they behave under infinite processes. First note that

$\mathbb{R}$ also inherits an *ordering* from $\mathbb{Q}$, which is naturally defined in terms of set containment.

Definition. We say $L_1 < L_2$ if $L_1 \subsetneqq L_2$.

An immediate consequence of this definition is a principle that was the main goal of Dedekind's definition of real numbers:

Least upper bound principle. *If X is any bounded set of real numbers then X has a least upper bound.*

Proof. Consider the lower Dedekind cuts L representing the numbers $x \in X$. Since X is bounded above, there is a rational number q greater than any member of any L. It follows that the union L^* of all the sets L is itself a lower Dedekind cut, defining some number x^*.

Clearly, each $L \subseteq L^*$, so $x \leq x^*$ for each $x \in X$. That is, x^* is *an* upper bound of X. And x^* is the *least* upper bound, because for any $y < x^*$ we have $y <$ some member of some L, and hence $y <$ some $x \in X$. □

The least upper bound principle was first stated by Bolzano (1817), before there was a definition of $\mathbb{R}$ precise enough to allow the principle to be proved. In the next section we will see that the least upper bound principle underlies many properties of $\mathbb{R}$ stating that a certain kind of infinite process is meaningful. These properties reflect the so-called *completeness* of $\mathbb{R}$.

A simple example is the infinite decimal 0.9999. . . . We take this expression to mean the least upper bound of the sequence 0.9, 0.99, 0.999, 0.9999, . . ., which exists by the least upper bound principle, and is necessarily 1. It similarly follows that any infinite decimal represents a well-defined real number.

Complex Numbers

From the foundational point of view the complex numbers involve no ideas beyond those used to construct $\mathbb{R}$, so we will not study them in detail. However, they are interesting historically as the first example where numbers were defined as ordered pairs. Hamilton (1835) defined the complex numbers $a + bi$ to be the ordered pairs $\langle a, b\rangle$ of real numbers, and he defined their sum and product by the rules

$$\langle a_1, b_1\rangle + \langle a_2, b_2\rangle = \langle a_1 + a_2, b_1 + b_2\rangle,$$
$$\langle a_1, b_1\rangle \cdot \langle a_2, b_2\rangle = \langle a_1 a_2 - b_1 b_2, a_1 b_2 + a_2 b_1\rangle.$$

The latter equation is motivated by the result

$$(a_1 + b_1 i)(a_2 + b_2 i) = a_1 a_2 - b_1 b_2 + (a_1 b_2 + a_2 b_1) i,$$

obtained by assuming that complex numbers $\mathbb{C}$ have the field properties and that $i^2 = -1$. With Hamilton's definitions of sum and product the field properties of the complex numbers are inherited from those of the real numbers.

2.3 COMPLETENESS PROPERTIES OF $\mathbb{R}$

The least upper bound property concerns bounded, but otherwise arbitrary, sets of real numbers. In the basic analysis studied in this book we are more likely to use bounded *sequences* of real numbers $x_0, x_1, x_2, \ldots$. Since the members of a sequence constitute a set, $\{x_0, x_1, x_2, \ldots\}$, it is also true that a bounded sequence of real numbers has a least upper bound. Since this special case of the least upper bound principle is particularly important, we will name it: the *sequential least upper bound principle*.

An even more special case is the *monotone convergence theorem*, stating that a bounded *nondecreasing* (or *nonincreasing*) sequence has a *limit*. We have not defined limit until now, but it is clear what the "limit" of a bounded nondecreasing sequence should be: its least upper bound.

Definition. A sequence $x_0, x_1, x_2, \ldots$ has *limit* l if, for each $\varepsilon > 0$, there is a natural number N such that

$$n > N \Rightarrow |x_n - l| < \varepsilon.$$

We write this relation as $\lim_{n\to\infty} x_n = l$, or sometimes $x_n \to l$ as $n \to \infty$, and also say that a sequence is *convergent* if it has a limit.

With this definition it is clear that if $x_0 \le x_1 \le x_2 \le \cdots$ and if l is the least upper bound of $\{x_0, x_1, x_2, \ldots\}$, then also $\lim_{n\to\infty} x_n = l$.

There is a similar story about *greatest lower bounds* and the limits of nonincreasing sequences. The greatest lower bound for any bounded set S exists, since it is the negative of the least upper bound of $\{-s : s \in S\}$, and the limit of a bounded nonincreasing sequence is clearly equal to its greatest lower bound.

Nested Sequences of Closed Intervals

Bounded monotonic sequences commonly occur in the setting of nested sequences of closed intervals. A closed interval is a set of the form

$$[a,b] = \{x \in \mathbb{R} : a \leq x \leq b\}.$$

A sequence $[a_0, b_0], [a_1, b_1], [a_2, b_2], \ldots$ of intervals is called *nested* if

$$[a_0, b_0] \supseteq [a_1, b_1] \supseteq [a_2, b_2] \supseteq \cdots .$$

Such sequences are implicit, for example, in infinite decimals, such as

$$x = 3.1415\cdots,$$

which encapsulates a sequence of nested intervals which "narrow down to" x, in the sense that x is the single common point of the nested sequence

$$[3,4] \supset [3.1,3.2] \supset [3.14,3.15] \supset [3.141,3.142] \supset [3.1415,3.1416] \supset \cdots .$$

The existence of a common point for an arbitrary nested sequence

$$[a_0, b_0] \supseteq [a_1, b_1] \supseteq [a_2, b_2] \supseteq \cdots$$

follows from the monotonicity of the sequences

$a_0 \leq a_1 \leq a_2 \leq \cdots$, which has least upper bound a, say, and
$b_0 \leq b_1 \leq b_2 \leq \cdots$, which has least upper bound b, say.

Since $a_n \leq a \leq b \leq b_n$ for each n, a and b belong to all the intervals $[a_n, b_n]$.

If, in addition, the intervals $[a_n, b_n]$ become arbitrarily small then we cannot have $a < b$ (because the intervals eventually have length $< b - a$). So in this case *the nested intervals have exactly one common point.* (This is so in the case of infinite decimals, because the length of the nth interval is 10^{-n}.)

We may call this property of $\mathbb{R}$ *nested interval completeness.*

The Cauchy Convergence Criterion

Being bounded and monotonic is a criterion for convergence of a sequence that does not require any mention of the limit of the sequence.

But monotonic sequences are very special. There is in fact a completely general convergence criterion that does not mention the limit, namely, the criterion of Cauchy (1821), chapter 6:

Cauchy convergence criterion. *A sequence $x_0, x_1, x_2, \ldots$ is convergent if and only if, for each $\varepsilon > 0$, there is an N such that*

$$m, n > N \Rightarrow |x_m - x_n| < \varepsilon.$$

Proof. If $x_0, x_1, x_2, \ldots$ is convergent, then it has a limit l and it follows from the definition of limit that, for each $\varepsilon > 0$, there is an N such that

$$n > N \Rightarrow |x_n - l| < \varepsilon/2.$$

It follows in particular that

$$m, n > N \Rightarrow |x_m - x_n| = |x_m - l - (x_n - l)| \leq |x_m - l| + |x_n - l| < \varepsilon/2 + \varepsilon/2 = \varepsilon,$$

as required.

Conversely, if $x_0, x_1, x_2, \ldots$ satisfies the Cauchy convergence criterion then the sequence is bounded, because the finite sequence $x_0, x_1, \ldots, x_N$ is bounded and the terms thereafter differ from x_N by at most ε. Then if we let

$$\begin{aligned} a_n &= \text{greatest lower bound of } \{x_n, x_{n+1}, x_{n+2}, \ldots\}, \\ b_n &= \text{least upper bound of } \{x_n, x_{n+1}, x_{n+2}, \ldots\}, \end{aligned}$$

we have nested intervals

$$[a_0, b_0] \supseteq [a_1, b_1] \supseteq [a_2, b_2] \supseteq \cdots,$$

which become arbitrarily small by the Cauchy criterion, and hence contain a single common point, l. It follows that, for any $\varepsilon > 0$, there is an N such that

$$n > N \Rightarrow |x_n - l| < \varepsilon,$$

and hence $x_0, x_1, x_2, \ldots$ converges to l. □

This theorem gives another common way to express completeness of $\mathbb{R}$: *every sequence satisfying the Cauchy criterion has a limit.*

2.4 FUNCTIONS AND SETS

We have now seen how integer and rational numbers reduce to natural numbers, and how real numbers reduce to sets of rational numbers (and hence to sets of natural numbers). The next objects to arithmetize are functions. In fact we have already considered certain kinds of functions.

The *sequence* $x_0, x_1, x_2, \ldots$ is a function f on the natural numbers with real number values, namely

$$f(n) = x_n.$$

This function can in turn be viewed as the *set of ordered pairs*

$$f = \{\langle 0, f(0)\rangle, \langle 1, f(1)\rangle, \langle 2, f(2)\rangle, \ldots\}.$$

Indeed, any function f with domain D can be viewed as the set

$$\{\langle x, y\rangle : x \in D \text{ and } f(x) = y\}.$$

This shows how functions may be reduced to sets, but it also adds to the plethora of ordered pairs we have already used to define integers and rational numbers. To achieve the ultimate goal of arithmetization—reducing all concepts of analysis to natural numbers and sets of natural numbers—we need a way to encode pairs of numbers by numbers.

Pairing Functions

A function that maps distinct ordered pairs of numbers to distinct numbers is called a *pairing function*. We begin by discussing the most important example: pairing functions on the set $\mathbb{N}$ of natural numbers. Figure 2.1 shows an arrangement of the set $\mathbb{N} \times \mathbb{N}$ of ordered pairs $\langle m, n\rangle$ of natural numbers that makes it clear that a pairing function exists. In fact it shows a *bijection* between $\mathbb{N} \times \mathbb{N}$ and $\mathbb{N}$.

If we order the diagonals in $\mathbb{N} \times \mathbb{N}$ from left to right and pairs along each diagonal from top to bottom, as shown, then $\langle m, n\rangle$ is the mth element on the $(m+n)$th diagonal, so it occurs at position

$$m + (0 + 1 + 2 + \cdots + (m+n)) = m + \frac{(m+n)(m+n+1)}{2}.$$

For example, $\langle 0, 0\rangle$ is at position 0 and $\langle 1, 2\rangle$ at position $1 + \frac{3\times 4}{2} = 1 + 6 = 7$.

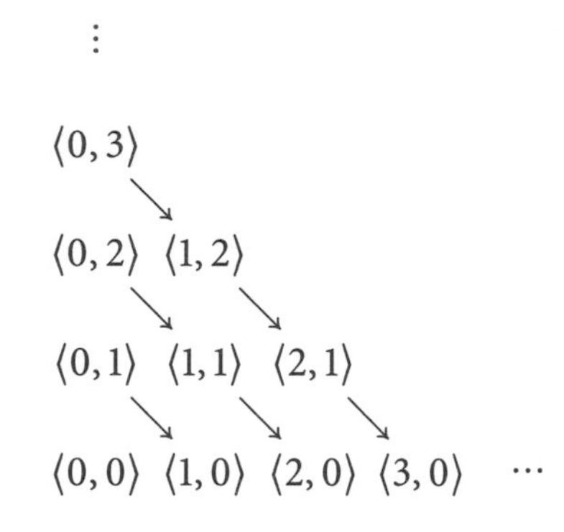

Figure 2.1 : Listing the ordered pairs of natural numbers

Thus the function

$$P(m,n) = m + \frac{(m+n)(m+n+1)}{2}$$

is a pairing function for $\mathbb{N} \times \mathbb{N}$. This pairing function is algebraically simple,[1] hence easily handled in the formal language of Peano arithmetic we introduce in section 2.7.

Using it, we can encode integers and rationals by natural numbers, and hence we can encode sets of rationals (and thereby real numbers) by sets of natural numbers. Thus we begin to see how analysis can be arithmetized, though there is reason to doubt that arithmetization can give a good account of *sets* of real numbers, as we will see in section 2.9. However, we can handle *sequences* of real numbers.

Encoding Sequences and Certain Other Functions

Given that each real x_n can be encoded by a set $X_n \subseteq \mathbb{N}$, we can encode the sequence $\{\langle n, x_n \rangle : n \in \mathbb{N}\}$ by the set of pairs

$$\{\langle n, k \rangle : k \in X_n \text{ and } n \in \mathbb{N}\},$$

and hence by the set of natural numbers

$$X = \{P(n,k) : k \in X_n \text{ and } n \in \mathbb{N}\}.$$

[1]This function is essentially due to Cantor (1895), §6, who introduced the corresponding function for *positive* integers. Surprisingly, it is the *only known quadratic* bijection from $\mathbb{N} \times \mathbb{N}$ to $\mathbb{N}$, apart from the one obtained by exchanging m and n. Pólya and Fueter (1923) proved that $P(m,n)$ and $P(n,m)$ are the only quadratic bijections that map $\langle 0, 0 \rangle$ to 0. I learned this from the book Smoryński (1991), p. 24.

Thus a sequence of real numbers can be encoded by a set $X \subseteq \mathbb{N}$ because it is a function with domain $\mathbb{N}$. The same idea works for any function whose domain can be *encoded by* $\mathbb{N}$, such as a function with domain $\mathbb{Q}$. In the next section we will see that this fact allows us to encode any *continuous* function on the real numbers by a set of natural numbers.

2.5 CONTINUOUS FUNCTIONS

The standard way to define continuous real functions, which goes back to Cauchy (1821), is as follows. It formalizes the idea that $f(x)$ "approaches" $f(a)$ as x "approaches" a.

Definitions. A real-valued function f is called *continuous at* $a \in \mathbb{R}$ if, for each $\varepsilon > 0$, there is a $\delta > 0$ such that

$$|x-a| < \delta \Rightarrow |f(x) - f(a)| < \varepsilon.$$

The function f is *continuous on* a set $S \subseteq \mathbb{R}$ if f is continuous at each $a \in S$.

A consequence of this definition is that a continuous function behaves in the expected way on convergent sequences.

Sequential continuity. *If f is continuous at $x = a$ and defined at points $a_0, a_1, a_2, \ldots$ with $\lim_{n\to\infty} a_n = a$, then $f(a) = \lim_{n\to\infty} f(a_n)$.*

Proof. By definition of continuity, for each $\varepsilon > 0$ there is a $\delta > 0$ such that

$$|x-a| < \delta \Rightarrow |f(x) - f(a)| < \varepsilon.$$

Also, since $\lim_{n\to\infty} a_n = a$, for each $\delta > 0$ there is an N such that

$$\begin{aligned} n > N &\Rightarrow |a_n - a| < \delta \\ &\Rightarrow |f(a_n) - f(a)| < \varepsilon \quad \text{by continuity at } x = a. \end{aligned}$$

And this says $\lim_{n\to\infty} f(a_n) = f(a)$ □

Now, since each real number a is (by the definition in section 2.2) the least upper bound of a set L of rational numbers, it is the limit of a sequence $a_0, a_1, a_2, \ldots$ of rational numbers. (Obtained, for example, by choosing each a_n from L at distance less than $1/n$ from a.) Therefore, each function f continuous at a and defined within some radius δ of a is defined on some sequence of rational numbers $a_0, a_1, a_2, \ldots$ with limit a. Thus it follows from sequential continuity that

$$f(a) = \lim_{n\to\infty} f(a_n).$$

In other words, *each value $f(a)$ of f is determined by the values of f at rational points.*

It follows, by the remark at the end of the previous section, that *each continuous function on $\mathbb{R}$ (or on an interval of $\mathbb{R}$) may be encoded by a set of natural numbers*, and hence that the arithmetization project extends at least as far as the continuous functions. This remarkable result is due to Borel (1898), p. 109, and it follows that each continuous function may be encoded by a real number. Is it not surprising that each continuous function, no matter how complicated, can be captured by a *point* on the number line?

Encoding Continuous Functions by Rational Intervals

There is a more direct way to define continuous functions, introduced by Hausdorff (1914), which leads to rather more natural encoding of them.

Definitions. An *open interval* of $\mathbb{R}$ is a set of the form

$$(a, b) = \{x \in \mathbb{R} : a < x < b\}.$$

An *open set* (in $\mathbb{R}$) is an arbitrary union of open intervals.

The characteristic property of an open set U is that any $x \in U$ is "properly inside" U in the sense that, for some $\delta > 0$, all points within distance δ of x are also in U. This leads to Hausdorff's characterization of continuous functions, in which we use the notation $f((c, d))$ to denote $\{f(x) : x \in (c, d)\}$.

Hausdorff characterization of continuity. *A real-valued function f on an open set U is continuous if and only if, for each value $f(x)$ and each (a, b) including $f(x)$, there is an open interval (c, d) including x such that $f((c, d)) \subseteq (a, b)$.*

Proof. Since f is continuous, for each $x \in U$ and each $\varepsilon > 0$ there is a $\delta > 0$ such that

$$|x - x'| < \delta \Rightarrow |f(x) - f(x')| < \varepsilon.$$

In particular, if $f(x) \in (a, b)$ and $\varepsilon = \min(f(x) - a, b - f(x))$ there is a $\delta > 0$ such that $f(x') \in (a, b)$ for $x' \in (x - \delta, x + \delta) \subseteq U$. So if we set $(x - \delta, x + \delta) = (c, d)$ we have

$$f((c, d)) \subseteq (a, b).$$

Conversely, if for each (a, b) including $f(x)$ there is a (c, d) including x with $f((c, d)) \subseteq (a, b)$ then, for each $\varepsilon > 0$, there is (c, d) including x with $f((c, d)) \subseteq (f(x) - \varepsilon, f(x) + \varepsilon)$.

Consequently, if $\delta = \min(x - c, d - x)$ we have

$$|x - x'| < \delta \Rightarrow |f(x) - f(x')| < \varepsilon,$$

so f is continuous at x for each $x \in U$, and hence f is continuous *on* U. □

It follows from the Hausdorff characterization that we can encode any f continuous on an open set U by the pairs of *rational* intervals $\langle (a, b), (c, d) \rangle$ such that $f((c, d)) \subseteq (a, b)$. This is because x and $f(x)$ are determined by the rational intervals containing them, and if $f((c, d))$ is contained in a rational interval (a, b) we can ensure that (c, d) is also rational, by making it slightly smaller, if necessary.

By suitable use of pairing functions we can encode the set of ordered pairs $\langle (a, b), (c, d) \rangle$, and hence the function f itself, by a set of natural numbers. Thus we recover the result found in the first part of this section.

2.6 THE PEANO AXIOMS

In the preceding sections of this chapter we have outlined how the basic concepts of analysis, from the real numbers to continuous functions, may be derived from the natural numbers and sets of natural numbers. Now it is time to lay the foundations of the natural numbers themselves, and their accompanying operations of addition and multiplication.

The bedrock concept of this foundation is *induction*, first recognized by Grassmann (1861) as the basis for *defining* addition and multiplication, and for *proving* their algebraic properties such as $a + b = b + a$. Grassmann's ideas were encapsulated in an axiom system for the natural numbers by Peano (1889), and this system is now known as *Peano arithmetic* (PA).

Over the years, the statement of the Peano axioms has varied slightly. In Peano's original system, 1 was taken as the least number; we take 0 to be least. There is also some variation in the induction axiom. It was originally stated as property of arbitrary sets of natural numbers but we now restrict it to properties definable in the language of PA, as explained below. One reason for the restriction is that we wish to separate the concept of set of natural numbers from the concept of natural number, so we no longer include the concept of set in the axioms of PA.

The proper place for axioms about sets is in axiom systems for analysis, as we will see at length in chapters 6 and 7.

Axioms for Successor

The first two Peano axioms express our intuition that the natural numbers are generated from 0 by applying the *successor function* S:

1. For all n, $0 \neq S(n)$.
2. For all m and n, $S(m) \neq S(n) \Rightarrow m \neq n$.

From these, we deduce $0 \neq S(0)$, then $S(0) \neq SS(0)$, $SS(0) \neq SSS(0)$, and so on. In this way we can prove that

$$0, \quad S(0), \quad SS(0), \quad SSS(0), \quad \cdots$$

(the terms denoting the natural numbers) are all different from each other.

Axioms for Sum and Product

The next two axioms implicitly define the functions $+$ and $\cdot$ by induction (also called "definition by recursion" or "recursive definition").

3. For all m and n, $m + 0 = m$ and $m + S(n) = S(m + n)$.
4. For all m and n, $m \cdot 0 = 0$ and $m \cdot S(n) = m \cdot n + m$.

These axioms define $+$ and $\cdot$ for all the terms $0, S(0), SS(0), SSS(0), \ldots$ that we interpret as the natural numbers. Axiom 3 defines $m + k$ for all m and for $k = 0$; then it defines $m + k$ for $k = S(n)$, provided that $m + n$ has already been defined. In fact from Axiom 3 we can prove all particular facts about sums of natural numbers, such as

$$SS(0) + SS(0) = SSSS(0)$$

(the equation we normally write as $2 + 2 = 4$).

However, we are not yet able to prove *general* facts about addition, such as $a + b = b + a$ for all a and b.

Axiom 4 defines $m \cdot k$ for all m and for $k = 0$; then it defines $m \cdot k$ for $k = S(n)$, provided that $m \cdot n$ and the + function have already been defined. For Axioms 3 and 4 we can prove all particular facts about sum and product of natural numbers, such as

$$SS(0) \cdot (S(0) + SSS(0)) = SSSSSSSS(0)$$

(the equation we normally write as $2(1 + 3) = 8$).

But, as before, we cannot prove general facts, such as $a \cdot (b + c) = a \cdot b + a \cdot c$ for all a, b, and c. These facts do not follow from Axioms 1–4, as can be shown by concocting a model of Axioms 1–4 that includes some alien objects (unequal to 0, S(0), SS(0), ...) with peculiar sums and products.

Axioms 1–4 are a very concise way to encapsulate all particular facts about sums and products of natural numbers—they cover something like "elementary school arithmetic." But to capture the "higher arithmetic" of *general* facts about numbers we need to formalize the principle of induction that we used informally to see what follows from Axioms 1–4. It is induction that enables us to prove the algebraic properties underlying the field properties of $\mathbb{Q}$ and $\mathbb{R}$.

Induction

Induction is the principle that allows us to conclude that a property $\varphi(n)$ holds for all n once we have proved

- $\varphi(0)$ holds (the "base step");
- for all n, if $\varphi(n)$ holds then $\varphi(S(n))$ holds (the "induction step").

Using the symbol $\forall n$ to denote "for all n" we express this principle by the axiom

5. $[\varphi(0) \text{ and } \forall n(\varphi(n) \Rightarrow \varphi(S(n)))] \Rightarrow \forall n \varphi(n)$.

Axiom 5 is known as the *induction axiom* or, more precisely, as the induction axiom *schema*. It really consists of infinitely many axioms, one for each property φ that can be written in the language of arithmetic. This language has variables for natural numbers, the function symbols S, +, and ·, the equality symbol =, parentheses, and logic symbols. The latter are symbols for "and," "or," "not," "implies," "for all," and "there exists." For more details, see the next section, which describes a classification of formulas in this language according to "quantifier complexity."

The induction axiom used by Peano is worth a brief mention here, because we will eventually use it in the system ACA_0 of analysis discussed in chapter 6. We call it *set variable induction* because it involves a set variable X and says $n \in X$ instead of $\varphi(n)$, namely

$$[0 \in X \text{ and } \forall n(n \in X \Rightarrow S(n) \in X)] \Rightarrow \forall n(n \in X).$$

Examples of Proofs by Induction

Proving even $a + b = b + a$ takes more space than we care to use here, so we will give only two simpler examples in the same vein.

Successor is +1. *For all natural numbers n, $S(n) = n + 1$.*

Proof. The number 1 is defined to be $S(0)$, so

$$\begin{aligned} n + 1 &= n + S(0) \\ &= S(n + 0) && \text{by definition of } +, \\ &= S(n) && \text{since } n + 0 = n \text{ by definition of } +. \end{aligned}$$ □

Commutativity of adding 1. *For all natural numbers n, $1 + n = n + 1$.*

Proof. Since $S(n) = n + 1$ by the previous proposition, it suffices to prove $S(n) = 1 + n$. We do this by induction on n.

For the base step $n = 0$ we have

$$S(0) = 1 = 1 + 0 \qquad \text{by definition of } +.$$

For the induction step we assume $S(k) = 1 + k$, so $k + 1 = 1 + k$, and consider $S(S(k))$:

$$\begin{aligned} S(S(k)) &= S(k + 1) && \text{by the previous proposition,} \\ &= S(1 + k) && \text{by the induction hypothesis,} \\ &= 1 + S(k) && \text{by definition of } +. \end{aligned}$$

This completes the induction step, so $S(n) = 1 + n$ for all natural numbers n. □

By further use of induction one can obtain more general results about natural numbers, such as $m + n = n + m$, $l + (m + n) = (l + m) + n$, $mn = nm$, and $l(m + n) = lm + ln$. Then, by introducing negative integers and rational numbers via ordered pairs as in section 2.1, we can prove all the ring properties of the integers and the field properties of the rationals. This was first done by Grassmann (1861), and is quite laborious. The property $mn = nm$ is Grassmann's theorem 72! However, each step is fairly routine, so we will skip the details.

2.7 THE LANGUAGE OF PA

The language of arithmetic, mentioned in the discussion of induction in the previous section, can be described more precisely as follows. Its symbols are:

Constant: 0

Variables: $a, b, c, \ldots$ (lowercase Roman letters)

Function symbols: $S, +, \cdot$ (for successor, sum, product)

Relation symbol: $=$

Logic symbols: $\wedge$ (and), $\vee$ (or), $\neg$ (not), $\Rightarrow$ (implies), $\forall$ (for all), $\exists$ (there is)

Parentheses: (,)

These symbols are combined by the following rules to build *terms*, *equations*, and *formulas*. The constant, variable, and function symbols are used to build terms: 0 or a variable is a term, and if t_1 and t_2 are terms then so are

$$S(t_1), \quad (t_1 + t_2), \quad \text{and} \quad (t_1 \cdot t_2).$$

In particular, the terms include the *numerals* $0, S(0), SS(0), \ldots$ for the natural numbers, which we will often abbreviate by their usual names $0, 1, 2, \ldots$. We need parentheses to distinguish between terms with potentially different meanings, such as

$$S((a + b)) \quad \text{and} \quad (S(a) + b).$$

But, as in ordinary mathematics, we omit parentheses when there is no risk of confusion. In particular, we write $SS(0)$ instead of $S(S(0))$, and so on.

From terms t_1, t_2 we can also build the *equation* $(t_1 = t_2)$ with the help of the equality symbol. An equation becomes a meaningful sentence when numerals are substituted for its variables. Equations are the simplest type of arithmetic formula; from them we build formulas in general with the help of logic symbols: the *connectives* or *Boolean operations* $\wedge, \vee, \neg$, and $\Rightarrow$ and the *quantifiers* $\forall$ and $\exists$. Thus if φ_1 and φ_2 are formulas then so are

$$(\varphi_1 \wedge \varphi_2), \quad (\varphi_1 \vee \varphi_2), \quad (\neg\varphi_1), \quad \forall x \varphi_1, \quad \exists x \varphi_1,$$

where x is a variable in φ_1 not already in the scope of $\forall$ or $\exists$.

The language of PA is capable of expressing all the usual sentences about natural numbers, and also all the usual relations between, and properties of, natural numbers. Some examples are:

1. $\forall n \neg(0 = S(n))$,
   which is the first Peano axiom, stating that 0 is not a successor.
2. $\exists l(l + m = n)$,
   which says that $m \leq n$.
3. $\exists l(l \cdot m = n)$,
   which says that m divides n.

Examples 2 and 3 show that the relations "$m \leq n$" and "m divides n" are definable in the language of PA, so we can use these relations (to help readability) on the understanding that they abbreviate the formulas in examples 2 and 3. For example, using the "divides" abbreviation and the abbreviation 1 for $S(0)$ we can define the property "p is prime" by the formula

4. $\forall l(l \text{ divides } p \Rightarrow (l = 1 \vee l = p))$.

We can define the relation $P(x, y) = z$, where P denotes the pairing function found in section 2.4, by the equation

5. $2 \cdot z = 2 \cdot x + (x + y) \cdot (x + y + 1)$,

where 1 abbreviates $S(0)$ and 2 abbreviates $SS(0)$. Also, if P_1 and P_2 are the *projection functions* that recover $x = P_1(z)$ and $y = P_2(z)$ from $z = P(x, y)$ then $x = P_1(z)$ means $\exists y(P(x, y) = z)$ and $y = P_2(z)$ means $\exists x(P(x, y) = z)$, so:

6. $\exists y(2 \cdot z = 2 \cdot x + (x + y)(x + y + 1))$
   expresses the relation $x = P_1(z)$ and
7. $\exists x(2 \cdot z = 2 \cdot x + (x + y)(x + y + 1))$
   expresses the relation $y = P_2(z)$.

Simplification of Connectives

The Boolean operation symbols $\wedge, \vee, \neg, \Rightarrow$ are used because they correspond to the words "and," "or," "not," and "implies" used in natural language, hence they make it easy to move between natural language and the language of PA. However, it is possible, and sometimes convenient, to work with a smaller set of connectives.

For example, we can drop the connective $\Rightarrow$ because

$$\varphi_1 \Rightarrow \varphi_2 \text{ is logically equivalent to } (\neg\varphi_1) \vee \varphi_2.$$

Also, we can drop either of $\wedge, \vee$ in favor of the other because

$$\varphi_1 \wedge \varphi_2 \text{ is logically equivalent to } \neg((\neg\varphi_1) \vee (\neg\varphi_2)), \text{ and}$$
$$\varphi_1 \vee \varphi_2 \text{ is logically equivalent to } \neg((\neg\varphi_1) \wedge (\neg\varphi_2)).$$

Thus it suffices to use just the connectives $\vee, \neg$.

Prenex Form

The simplification just achieved, reducing to the connectives $\vee, \neg$, leads to a more important simplification in the use of quantifiers: the so-called *prenex form* in which all quantifiers are at the front of the formula. Prenex form is achieved by systematically applying the following equivalences (where we write $\Leftrightarrow$ to denote logical equivalence) to move quantifiers to the left of connectives:

$$\begin{aligned} \neg\forall x\varphi &\Leftrightarrow \exists x\neg\varphi \\ \neg\exists x\varphi &\Leftrightarrow \forall x\neg\varphi \\ \varphi_1 \vee \forall x\, \varphi_2(x) &\Leftrightarrow \forall y(\varphi_1 \vee \varphi_2(y)) \\ \varphi_1 \vee \exists x\, \varphi_2(x) &\Leftrightarrow \exists y(\varphi_1 \vee \varphi_2(y)) \end{aligned}$$

In the latter two equivalences we rename the quantified variable x in φ_2, if necessary, by a variable y not occurring in φ_1.

2.8 ARITHMETICALLY DEFINABLE SETS

From now on we call the properties definable in PA, and the corresponding sets, *arithmetically definable*. Using the equivalences above, we can reduce any arithmetically definable property $\alpha(u)$ to one defined by a formula in prenex form:

$$Q_1x_1Q_2x_2\cdots Q_nx_n\ \varphi(x_1,x_2,\ldots,x_n,u),$$

where $Q_1,Q_2,\ldots,Q_n$ are quantifiers $\forall$ or $\exists$, and φ is *quantifier-free*. That is, φ consists of a series of equations between terms linked by the connectives $\vee$ and $\neg$ (also known as a *Boolean combination* of equations). And terms, as we explained in the previous section, are built from the variables $x_1,x_2,\ldots,x_n,u$ and the constant 0 by means of the S, $+$, and $\cdot$ functions.

At the cost of complicating terms by inclusion of the projection functions P_1 and P_2, we can reduce any two adjacent quantifiers of the same type to a single one, because

$$\forall x\forall y\ \varphi(x,y)\Leftrightarrow\forall z\ \varphi(P_1(z),P_2(z)),\ \text{and}$$
$$\exists x\exists y\ \varphi(x,y)\Leftrightarrow\exists z\ \varphi(P_1(z),P_2(z)).$$

By such reductions we eventually arrive at a prenex formula in which the quantifiers *alternate*, either

$$\forall z_1\exists z_2\cdots\ \psi(z_1,z_2,\ldots,z_m,u)$$

or

$$\exists z_1\forall z_2\cdots\ \psi(z_1,z_2,\ldots,z_m,u),$$

where ψ is a quantifier-free combination of equations between terms built from $z_1,z_2,\ldots,z_m,u$ using the functions $S,+,\cdot,P_1$, and P_2. The first type, in which there are m alternating quantifiers beginning with $\forall$, is called a Π^0_m formula. The second is called a Σ^0_m formula.[2]

It turns out that m is a good measure of the complexity of an arithmetical property $\alpha(u)$. In particular, a Σ^0_1 property is one that is "computably enumerable" in a sense we will explain in the next subsection.

Σ^0_1 *Properties*

It follows from the definition of Σ^0_m above that a property $\alpha(u)$ is Σ^0_1 if there is a quantifier-free formula $\psi(z,u)$ such that

$$\alpha(u)\Leftrightarrow\exists z\ \psi(z,u).$$

[2]The reasons for the notation include: Π because the $\forall$ quantifier is like a logical "product," Σ because $\exists$ is like a logical "sum," and the superscript 0 indicates that the quantifiers are over the objects of lowest type, the natural numbers. This leaves the option of a possible superscript 1 when quantifers are over *sets* of natural numbers.

The formula $\psi(z, u)$ is a Boolean combination of equations $t_1 = t_2$ between terms built from the variables u, z and the constant 0 by means of the functions $S, +, \cdot, P_1$, and P_2. These functions are obviously *computable*, so for any values of the variables u, z we can compute whether each equation $t_1 = t_2$ is true or not.

Moreover, $\psi(z, u)$ is a combination of these equations by the connectives $\vee$ and $\neg$. So, given the truth values ("true" or "false") of the equations, we can compute the truth value of the combination $\psi(z, u)$ with the help of the so-called *truth tables* for $\vee$ and $\neg$: $e_1 \vee e_2$ is true just in case one of e_1, e_2 is true, $\neg e$ is true just in case e is false.

Thus, for any given values of u, z, we can compute whether $\psi(z, u)$ is true or false. By systematically trying all pairs $\langle u, z\rangle$ we can eventually find each u for which $\exists z\ \psi(z, u)$. We can therefore make a list of all such u, which is why we say that the set of u such that $\exists z\ \psi(z, u)$ is *computably enumerable*.

Note that we do not claim to be able to compute, for each u, whether $\exists z\ \psi(z, u)$ is *true*. We claim only that, if it *is* true, we will eventually find out. If $\exists z\ \psi(z, u)$ is false we will search endlessly for a z such that $\psi(z, u)$, and may never know that our search is in vain. Indeed we will show, in section 4.2, that there are Σ_1^0 properties $\exists z\ \psi(z, u)$ for which there is no algorithm to compute the truth value of $\exists z\ \psi(z, u)$ for each u.[3]

The concept of a computably enumerable property obviously depends on the definition of "computable," a concept we study more deeply in chapters 4 and 5. However, we can reveal that the concept of computable enumerability *coincides* with the concept of a Σ_1^0 property, in the following precise sense.

If we define a Σ_1^0 property to be one defined by a formula

$$\exists x_1 \exists x_2 \cdots \exists x_n\ \psi(x_1, x_2, \ldots, x_n, u), \qquad (*)$$

where ψ is a Boolean combination of equations between terms built from variables, 0, and the $S, +$, and $\cdot$ functions, then such a property is clearly computably enumerable, by an argument like that above. Conversely, once the concept of "computable" is precisely defined, it can be shown that any computably enumerable relation is of the form (*). In fact, it was

[3]All mathematicians know computable properties $\psi(z, u)$ with no *known* algorithm to compute the truth value of $\exists z\ \psi(z, u)$. One is the property "u consecutive zeros occur before the zth decimal place of π." But it is a different matter to prove that no algorithm exists.

shown by Matijasevič (1971) that $\psi(x_1, x_2, \ldots, x_n, u)$ can be a *single equation* $t_1 = t_2$, where t_1 and t_2 are terms built from the S, $+$, and $\cdot$ functions. In other words, t_1 and t_2 are *polynomials* in the variables $x_1, x_2, \ldots, x_n, u$.

Thus computably enumerable properties, and hence all Σ^0_1 properties, are in fact Σ^0_1 of a particularly simple form.

2.9 LIMITS OF ARITHMETIZATION

In section 2.4 we raised a doubt whether arithmetization can give a good account of *sets* of real numbers—in contrast to *sequences* of real numbers, which can be encoded by sets of natural numbers. In this section we will explain why sets of real numbers cannot all be encoded by sets of natural numbers. This puts arbitrary sets of real numbers beyond the scope of arithmetization as it is normally understood.

At the same time we will see that sets of natural numbers cannot all be encoded by natural numbers. This is why sets of natural numbers are viewed as objects different in type from the natural numbers themselves (and why the arithmetization project for analysis needs *sets* of natural numbers as well as natural numbers). An equivalent statement is that the *real* numbers (which correspond to sets of natural numbers) cannot all be encoded by natural numbers. This limits how many real numbers we can *define*, because definitions are finite strings of symbols, which can be encoded by natural numbers.

The following theorem of Cantor (1891) explains both of the facts above, and much more.

Cantor's theorem. *For each set S there is no one-to-one correspondence between the elements of S and the subsets of S.*

Proof. Suppose that, for each $x \in S$, there corresponds a subset $S_x \subseteq S$. It suffices to show that the subsets S_x do not include all the subsets of S.

Indeed, they do not include the subset

$$X = \{x \in S : x \notin S_x\},$$

because X differs from S_x with respect to the element x: $x \in X \Leftrightarrow x \notin S_x$.
□

Obviously, there is a one-to-one correspondence between the elements of S and *certain* subsets of S—for example, we can let the element x correspond to the subset $\{x\}$—so Cantor's theorem essentially says that *any set has more subsets than elements.* In particular, there are more sets of natural numbers than natural numbers, and more sets of real numbers than real numbers. This is why a set of natural numbers cannot in general be encoded by a natural number, and why a set of real numbers cannot in general be encoded by a real number (or by a set of natural numbers).

The Diagonal Argument

The devastatingly simple argument in Cantor's theorem, called the *diagonal* argument, can be adapted to many situations. It shows, for example, that if we have a defined sequence $S_0, S_1, S_2, \ldots$ of sets of natural numbers, then we can explicitly define a set X *not* in the sequence by

$$n \in X \Leftrightarrow n \notin S_n.$$

In particular, if the sequence is arithmetically definable—in the sense that the relation $m \in S_n$ is arithmetically definable—then D is itself an arithmetically definable set. We conclude immediately that *it is impossible to arithmetically define the sequence of all arithmetically definable sets*—even though we can certainly compute a list of all the formulas φ that define these sets. The catch is that the relation "$\varphi(m)$ holds for the nth formula φ on the list" is not arithmetically definable. This follows from a further refinement of the diagonal argument that we will see in section 4.3.

There we will show that the Σ_1^0 sets can be arranged in a sequence for which the diagonal set X (necessarily not Σ_1^0) is Π_1^0. In the language of computability, there is a computably enumerable set whose complement is not computably enumerable. The generalization of this result to any number of quantifiers is that, for each k, there is a Π_k^0 set that is not in Σ_k^0. It follows that Σ_{k+1}^0, which includes Π_k^0, is a larger class of sets than Σ_k^0.

Now if $m \in S_n$ is an arithmetically definable relation, in Σ_k^0 say, then $S_0, S_1, S_2, \ldots$ are all in Σ_k^0. Thus an arithmetically definable sequence fails to include all arithmetically definable sets because Σ_k^0 fails to include Σ_{k+1}^0.

Definability and Computability

The result just described is typical of the way our ability to define sets falls short of the totality of all sets. It means that there is no well-defined class of "definable sets"—rather, we have to *choose* classes of sets that are big enough for the purposes we have in mind.

In chapters 6 and 7 we will study the two most obvious choices: the arithmetically definable sets and the computable sets. A system ACA_0 of analysis based on arithmetically definable sets is big enough to prove all the theorems of basic analysis—the intermediate and extreme value theorems, Heine-Borel and Bolzano-Weierstrass theorems—and even some theorems thought to be quite hard, such as the Brouwer fixed point and Jordan curve theorems. A system RCA_0 based on computable sets is not as strong, but still useful. The only theorem on the above list that RCA_0 can prove is the intermediate value theorem. However, RCA_0 is strong enough to prove *equivalences* between many theorems that it cannot prove outright.

This makes RCA_0 a good base theory for finding which theorems of analysis are really equivalent to each other, and which are stronger than others. A surprising outcome of this investigation, which will unfold in chapters 6 and 7, is that most of the basic theorems of analysis fall into just three levels of "strength."

CHAPTER 3

■■■■■

Classical Analysis

In the previous chapter we established that many of the basic objects of analysis, such as real numbers, sequences, infinite series, and continuous functions, can be *arithmetized*. That is, they can be defined in terms of, or encoded by, natural numbers and sets of natural numbers. This discovery paves the way for *axiom systems* of analysis, based on the system PA of Peano arithmetic.

In the present chapter, we explore the basic concepts that arise when real numbers and continuous functions are studied, particularly the limit concept and its use in proving properties of continuous functions. We give proofs of the Bolzano-Weierstrass and Heine-Borel theorems, and the intermediate and extreme value theorems for continuous functions. Also we use the Heine-Borel theorem to prove *uniform* continuity of continuous functions on closed intervals, and its consequence that any continuous function is *Riemann integrable* on closed intervals.

In several of these proofs there is a construction by "infinite bisection," which can be recast as an argument about *binary trees*. The role of trees in analysis will be explored more fully in chapter 7, but in this chapter we use it to construct an object that will be important in that chapter—the so-called *Cantor set*.

3.1 LIMITS

Limits of Sequences

Analysis is fundamentally concerned with the outcomes of infinite processes on real numbers, or *limits*. For example, the following equations express the real numbers 1/3, $\sqrt{2}$, and π as the outcomes of certain infinite processes.

$$\frac{1}{3} = 0.333333\cdots$$

$$\sqrt{2} = 1 + \cfrac{1}{2 + \cfrac{1}{2 + \cfrac{1}{2 + \cfrac{1}{\ddots}}}}$$

$$\frac{\pi}{4} = 1 - \frac{1}{3} + \frac{1}{5} - \frac{1}{7} + \frac{1}{9} - \cdots.$$

In each case the right-hand side arises from a process producing an infinite sequence of rational numbers:

- The infinite decimal arises from the sequence of finite decimals
$$0.3, \quad 0.33, \quad 0.333, \quad \ldots.$$
- The infinite continued fraction arises from the sequence of finite fractions
$$1, \quad 1 + \frac{1}{2}, \quad 1 + \frac{1}{2 + \frac{1}{2}}, \quad \ldots.$$
- The infinite sum arises from the sequence of partial sums
$$1, \quad 1 - \frac{1}{3}, \quad 1 - \frac{1}{3} + \frac{1}{5}, \quad \ldots.$$

And the left-hand side is the *limit* of the sequence, as defined in section 2.3.

Limits of Functions

A sequence $a_1, a_2, a_3, \ldots$ may be seen as a function f on the positive integers,

$$f(n) = a_n,$$

so we can view its limit l as the limit of $f(n)$ as n tends to infinity. We also write $f(n) \to l$ as $n \to \infty$, as mentioned in section 2.3.

More generally, we can define limits of a real-valued function f of a real variable x (a special case of this occurred in the definition of continuity at a point in section 2.5).

Definitions. If f is a real-valued function, defined on some subset of $\mathbb{R}$,

1. $f(x) \to l$ as $x \to a$ if, for each $\varepsilon > 0$, there is a $\delta > 0$ such that

$$|x - a| < \delta \Rightarrow |f(x) - l| < \varepsilon.$$

2. $f(x) \to l$ as $x \to \infty$ if, for each $\varepsilon > 0$, there is an $N > 0$ such that

$$x > N \Rightarrow |f(x) - l| < \varepsilon.$$

Limit Points of a Set

Definition. A point l is called a *limit point* of a set $S \subseteq \mathbb{R}$ if, for each $\varepsilon > 0$, there are points of S other than l in the interval $(l - \varepsilon, l + \varepsilon)$. We also express the defining condition by saying "each *neighborhood* of l contains points of S other than l."

An important example is the set $S = \mathbb{Q}$ of rational numbers. Here, every real number x is a limit point of S, since each neighborhood of x contains members of the lower Dedekind cut for x, which are rational.

You may wonder: do we really need the concept of limit point? Might not every limit point be the limit of a sequence of points drawn from S? Well, yes, but we will see in section 3.4 that even the existence of limit points is a weighty question, which brings to light some important issues in the foundations of analysis.

3.2 ALGEBRAIC PROPERTIES OF LIMITS

To avoid many tedious and complicated calculations of "δ for given ε" we prove that the sum, difference, product, and quotient of convergent sequences are themselves convergent, and that they converge to the expected values.

Algebra of limits. *If $a_n \to a$ and $b_n \to b$ as $n \to \infty$ then also*

$$a_n + b_n \to a + b \tag{1}$$

$$a_n - b_n \to a - b \tag{2}$$

$$a_n \cdot b_n \to a \cdot b \tag{3}$$

$$a_n / b_n \to a/b \quad \text{if} \quad b \neq 0. \tag{4}$$

Proof. For (1) we have to make $a_n + b_n$ within ε of $a + b$ for n greater than a suitably chosen N. Given that $a_n \to a$ and $b_n \to b$ we can find positive integers A, B so that

$$n > A \Rightarrow |a_n - a| < \frac{\varepsilon}{2}$$

$$\text{and} \quad n > B \Rightarrow |b_n - b| < \frac{\varepsilon}{2}.$$

Then with $N = \max(A, B)$ we have $|a_n - a| < \varepsilon/2$ *and* $|b_n - b| < \varepsilon/2$ for $n > N$, so

$$\begin{aligned} |a_n + b_n - (a + b)| &\leq |(a_n - a) + (b_n - b)| \\ &\leq |a_n - a| + |b_n - b| \\ &\leq \frac{\varepsilon}{2} + \frac{\varepsilon}{2} = \varepsilon. \end{aligned}$$

Thus $a_n + b_n \to a + b$ as $n \to \infty$.

For (2) the argument is similar.

For (3) we want to make $a_n b_n - ab < \varepsilon$. To do this we use a trick, writing

$$a_n b_n - ab = a_n b_n - ab_n + ab_n - ab = b_n(a_n - a) + a(b_n - b).$$

Now we have to make $|a_n - a|$ and $|b_n - b|$ small enough to compensate for the factors b_n and a. By making $|b_n - b| < \varepsilon/2|a|$ (by choosing $n > B$ say) we can compensate for the factor a.

To compensate for the variable factor b_n we first choose n large enough to make $|b_n| < 2|b|$, which is possible because $b_n \to b$. Then we make n larger, if necessary, so as to make $|a_n - a| < \varepsilon/4|b|$ (by choosing $n > A$ say). Then for $n > N = \max(A, B)$ we have

$$\begin{aligned} |a_n b_n - ab| &= |b_n(a_n - a) + a(b_n - b)| \\ &\leq |b_n||a_n - a| + |a||b_n - b| \\ &\leq 2|b|\frac{\varepsilon}{4|b|} + |a|\frac{\varepsilon}{2|a|} \\ &= \frac{\varepsilon}{2} + \frac{\varepsilon}{2} = \varepsilon, \end{aligned}$$

as required to prove that $a_n b_n \to ab$.

For (4) we view a_n/b_n as the product of a_n and $1/b_n$. Then, by the result (3) we are reduced to proving that

$$\frac{1}{b_n} \to \frac{1}{b} \quad \text{as} \quad n \to \infty \quad \text{when} \quad b \neq 0.$$

To do this we have to make $\left|\frac{1}{b_n} - \frac{1}{b}\right| < \varepsilon$. Well,

$$\frac{1}{b_n} - \frac{1}{b} = \frac{b - b_n}{bb_n},$$

and we can make this smaller in absolute value than ε as follows. First make $|b_n| > |b|/2 > 0$, which is possible for sufficiently large n since $b_n \to b \neq 0$, and then choose n larger, if necessary (say $n > N$) so as to make $|b - b_n| < \varepsilon|b|^2/2$.

This gives

$$\left|\frac{1}{b_n} - \frac{1}{b}\right| = \left|\frac{b - b_n}{bb_n}\right| \leq \frac{\varepsilon|b|^2/2}{|b|^2/2} = \varepsilon,$$

as required. □

There are similar proofs that if $f(x) \to l$ and $g(x) \to m$ as $x \to c$ then also

1. $f(x) + g(x) \to l + m$,
2. $f(x) - g(x) \to l - m$,
3. $f(x) \cdot g(x) \to l \cdot m$,
4. $f(x)/g(x) \to l/m$, if $m \neq 0$.

3.3 CONTINUITY AND INTERMEDIATE VALUES

Our intuition of a continuous function is one whose graph is "unbroken": that is, a curve "without gaps" just as the number line $\mathbb{R}$ is without gaps. However, the usual definition of continuous function is in terms of the limit concept, and the idea of its graph having no gaps is captured by a theorem—the *intermediate value theorem*.

Recall from section 2.5 that a function f is *continuous at* c if $f(x) \to f(c)$ as $x \to c$; f is *continuous on* a set $S \subseteq \mathbb{R}$ if f is continuous at each point in S.

Intermediate value theorem. *If f is continuous on an interval $[a, b]$, with $f(a) < 0$ and $f(b) > 0$, then $f(c) = 0$ for some c in $[a, b]$.*

Before proving the theorem we remark that there is nothing special about the value 0 of f. A similar proof shows that f takes *every* value between $f(a)$ and $f(b)$—hence the name "intermediate value theorem." In fact, the endpoints a and b are unnecessarily general. We can take $a = 0$ and $b = 1$ without loss of generality, as we do in the proof below.

Proof. On exactly one half of the interval [0,1]—either $[0,1/2]$ or $[1/2,1]$—we have $f(x) \leq 0$ at one end of the subinterval and $f(x) \geq 0$ at the other. If $f(x) = 0$ at either end we are done. If not, we have an interval $[a_1, b_1]$ with $f(a_1) < 0$ and $f(b_1) > 0$ and we can repeat the process in $[a_1, b_1]$.

That is, if $f(x) \neq 0$ at the midpoint of $[a_1, b_1]$ we get exactly one half $[a_2, b_2]$ of $[a_1, b_1]$ for which $f(a_2) < 0$ and $f(b_2) > 0$. Proceeding in this way we either find a bisection point x, at some stage, where $f(x) = 0$, or else we obtain an infinite nested sequence of closed intervals

$$[a_1, b_1] \supset [a_2, b_2] \supset [a_3, b_3] \supset \cdots$$

such that $f(a_n) < 0$ and $f(b_n) > 0$ for each n. Also, since each interval is half the one before, they have a single common point, c, by nested interval completeness (section 2.3).

We then must have $f(c) = 0$. If $f(c) > 0$ then the continuity of f gives $f(a_n), f(b_n) > 0$ for any a_n, b_n sufficiently close to c, contrary to the construction of a_n, b_n. And $f(c) < 0$ is ruled out similarly.

Thus we either find a point where $f(x) = 0$ at some stage of the bisection process, or else we obtain such a point as the limit of intervals arising in the bisection process. □

Essentially the same proof as this was given by Cauchy (1821). An English translation of Cauchy's proof may be found in Bradley and Sandifer (2009), pages 309–311.

The Fundamental Theorem of Algebra

It follows from the algebraic properties of limits in the previous section, and the definition of continuity, that the sum, difference, and product of continuous functions are continuous. Add to these the easily checked results that constant functions and the identity function are continuous, and we can infer the *continuity of any polynomial function*

$$f(x) = a_n x^n + \cdots + a_1 x + a_0, \quad \text{where} \quad a_0, a_1, \ldots, a_n \in \mathbb{R}.$$

Notice also that for n odd, $f(x)$ will have the same sign as a_n for n large and positive, and the opposite sign for n large and negative. This is because $a_n x^n$ exceeds the sum of all the others in absolute value when x is sufficiently large, and $x^n > 0$ for $x > 0$ and $x^n < 0$ for $x < 0$.

Thus it follows from the intermediate value theorem that $f(x) = 0$ *for some x when f is an odd-degree polynomial with real coefficients.* In fact we have shown that any such equation has a *real* solution.

This is a special case of the *fundamental theorem of algebra* (FTA). The general case removes the restriction that n be odd, and allows the solution to be a complex number. Gauss (1816) gave a proof of FTA by reducing any polynomial equation with real coefficients, $f(x) = 0$, to one of odd degree by a purely algebraic (though complicated) argument. He was able to repeatedly divide the degree of the equation by 2—thus ultimately obtaining an odd-degree equation—with the help of quadratic equations (which is where complex solutions arise).

One such reduction may be found in Dawson (2015). From the foundational point of view, the subtlest part of Gauss's proof is the odd-degree case and its reliance on the intermediate value theorem. This is the part that rests upon the completeness of the real numbers. Indeed, it was the attempt of Bolzano (1817) to justify Gauss's proof—by proving the intermediate value theorem—that brought the issue of completeness to light. Bolzano appealed to completeness in terms of least upper bounds but, as the proof above shows, one may also appeal to the existence of a common point in each nested sequence of closed intervals.

3.4 THE BOLZANO-WEIERSTRASS THEOREM

It is clear that a set cannot have a limit point unless it is infinite, and that certain unbounded infinite sets (such as $\mathbb{N}$) do not have limit points either. However, finiteness and unboundedness are the only two obstructions to the existence of limit points.

Bolzano-Weierstrass theorem. *If S is an infinite set of points between real numbers a and b, then S has a limit point.*

Proof. Without loss of generality we can take $a = 0$ and $b = 1$, so S is a set of points in $[0,1]$. Since S is an infinite set, at least one half of $[0,1]$—either $[0,1/2]$ or $[1/2,1]$—contains infinitely many points of S.

We let $[a_1, b_1]$ be the leftmost half of $[0,1]$ that contains infinitely many points of S, and repeat the argument in $[a_1, b_1]$. This gives a half $[a_2, b_2]$ of $[a_1, b_1]$ that also contains infinitely many points of S. Continuing in this way we obtain an infinite nested sequence of closed intervals

$$[a_1, b_1] \supset [a_2, b_2] \supset [a_3, b_3] \supset \cdots,$$

each of which contains infinitely many members of S.

Since each interval is half the length of its predecessor, there is a single point c common to all the intervals $[a_n, b_n]$. Also, since these intervals become arbitrarily small, each ε-neighborhood of c contains some $[a_n, b_n]$, and hence infinitely many points of S.

Thus c is a limit point of the set S. □

Corollary (sequential Bolzano-Weierstrass). *A bounded infinite sequence* $x_1, x_2, x_3, \ldots$ *of reals contains a convergent subsequence* $x_{n_1}, x_{n_2}, x_{n_3}, \ldots$.

Proof. Let $S = \{x_1, x_2, x_3, \ldots\}$, and find an infinite nested sequence of closed intervals

$$[a_1, b_1] \supset [a_2, b_2] \supset [a_3, b_3] \supset \cdots,$$

as in the proof of the theorem. Now define a subsequence of $x_1, x_2, x_3, \ldots$ by

$$\begin{aligned} x_{n_1} &= x_1, \\ x_{n_k} &= \text{next term in the sequence, after } x_{n_{k-1}}, \text{ that is in } [a_k, b_k]. \end{aligned}$$

Since there are infinitely many terms of the sequence in $[a_k, b_k]$, x_{n_k} is always defined, so the subsequence is infinite. And it is convergent (to c) because its kth term lies in $[a_k, b_k]$, and hence within distance 2^{-k} of c. □

We prove the sequential Bolzano-Weierstrass theorem to get around the difficulty, noted in section 2.9, of arithmetizing the concept of *set* of real numbers. Notice that the sequence of intervals $[a_k, b_k]$ is *definable* (by induction) but not obviously *computable*, since a finite computation cannot test whether there are infinitely many terms of the sequence $x_1, x_2, x_3, \ldots$ in a given interval. This prevents us from computing the sequence of intervals $[a_k, b_k]$. This state of affairs is a clue that classical analysis sometimes requires non-computable processes, a fact that will gradually become clearer.

3.5 THE HEINE-BOREL THEOREM

The process of infinite bisection, used to "narrow the region of infinitude" in the proof of Bolzano-Weierstrass, can be used in other situations. In this section we study another situation of fundamental importance, in the Heine-Borel theorem. Yet despite its similar proof, Heine-Borel is weaker than Bolzano-Weierstrass in a subtle but precisely definable way (see chapter 7).

Heine-Borel theorem. *If S is an infinite set of open intervals that covers $[0,1]$, then some finite subset of S also covers $[0,1]$.*

Proof. Suppose on the contrary that no finite subset of S covers [0,1]. It follows that at least one half of [0,1]—namely, [0,1/2] or [1/2,1]—also cannot be covered by finitely many members of S.

Let $[a_1, b_1]$ be the leftmost such half and repeat the argument in $[a_1, b_1]$. This gives a half $[a_2, b_2]$ of $[a_1, b_1]$ that cannot be covered by finitely many members of S, and so on. In this way obtain an infinite nested sequence of closed intervals

$$[a_1, b_1] \supset [a_2, b_2] \supset [a_3, b_3] \supset \cdots,$$

none of which can be covered by finitely many members of S.

Since each $[a_n, b_n]$ is half the length of its predecessor, these intervals have a single common point c. But c lies inside some open interval I belonging to S, and therefore so does any sufficiently small $[a_n, b_n]$. This contradicts the conclusion that $[a_n, b_n]$ cannot be covered by finitely many members of S. So it was wrong to suppose that [0,1] cannot be covered by finitely many members of S. □

Corollary (sequential Heine-Borel). *If $I_1, I_2, I_3, \ldots$ is an infinite sequence of open intervals that covers $[0,1]$, then the finite sequence $I_1, I_2, \ldots, I_n$ also covers $[0,1]$ for some n.*

Proof. Let $S = \{I_1, I_2, I_3, \ldots\}$. Then it follows from the theorem that finitely many of the I_k cover [0,1]. And, for some n, these I_k are included in the sequence $I_1, I_2, \ldots, I_n$. □

The openness of the intervals I in S comes into play when we conclude from $c \in I$ that some $[a_n, b_n] \subset I$. In fact, it is possible to cover [0,1] by infinitely many *closed* intervals, no finite subset of which covers [0,1]. An example is the following sequence of closed intervals: [0,0] (which covers 0), followed by $[1/2,1], [1/4,1/2], [1/8,1/4], \ldots$ (which cover the remaining points).

Heine-Borel has many important consequences, some of which we will see in section 3.7. But while we are on a roll with infinite bisection constructions we will do one more of fundamental importance.

3.6 THE EXTREME VALUE THEOREM

We saw in section 3.3 that the graphs of continuous functions have "no gaps in the middle" in a certain sense (the intermediate value theorem). We now show that, on closed intervals, they also have "no gaps at the top or bottom." To be precise, we have:

Extreme value theorem. *If f is a continuous function on $[0,1]$ then f attains both a maximum and a minimum value on $[0,1]$.*

Proof. We first prove that f is *bounded* on [0,1]. Suppose on the contrary that f is unbounded on [0,1]; that is, f takes arbitrarily large positive or negative values.

In that case f is unbounded on some half of [0,1]. As usual, we let the leftmost such half be $[a_1, b_1]$, and repeat the argument ("narrowing towards a point of unboundedness"). This ultimately gives an infinite nested sequence of closed intervals

$$[a_1, b_1] \supset [a_2, b_2] \supset [a_3, b_3] \supset \cdots,$$

with f unbounded on each $[a_n, b_n]$. Since each $[a_n, b_n]$ is half the one before, there is a single point c common to all the $[a_n, b_n]$.

But since f is continuous, we can find an $[a_n, b_n]$ on which the values of f differ from $f(c)$ by less than a given $\varepsilon > 0$. This means that f is *bounded* on $[a_n, b_n]$, and we have a contradiction. This contradiction shows that f is indeed bounded on [0,1]. Therefore, by the completeness of $\mathbb{R}$, there is a least upper bound l to the values of f on [0,1].

If l is not a value of f then the function $\frac{1}{l-f(x)}$ is continuous and *un*bounded on [0,1], which we have just proved to be impossible. Therefore, l is in fact the maximum value of f on [0,1]. We can similarly show the existence of a minimum value. □

3.7 UNIFORM CONTINUITY

Recall from section 2.5 that if f is defined to be *continuous at* $x = c$ if, for each $\varepsilon > 0$, there is a δ such that

$$|x - c| < \delta \Rightarrow |f(x) - f(c)| < \varepsilon.$$

It follows that

$$x, x' \in (c - \delta, c + \delta) \Rightarrow |f(x) - f(x')| < 2\varepsilon$$

because

$$\begin{aligned} |f(x) - f(x')| &= |f(x) - f(c) + f(c) - f(x')| \\ &\leq |f(x) - f(c)| + |f(c) - f(x')| \\ &\leq \varepsilon + \varepsilon = 2\varepsilon. \end{aligned}$$

So, renaming 2ε as ε, we can rephrase the condition for continuity at $x = c$ as follows: for each $\varepsilon > 0$ there is a $\delta > 0$ such that

$$x, x' \in (c - \delta, c + \delta) \Rightarrow |f(x) - f(x')| < \varepsilon. \qquad (*)$$

The δ in this condition depends on c, so if f is continuous over a certain set S we have potentially varying values $\delta(c)$ for the same ε as c varies over S.

If we can find a δ satisfying (*) for all c in S then we have what is called *uniform continuity* on S. Replacing the δ in (*) by $\delta/2$ we get a more concisely stated condition:

Definition. A function f is called *uniformly continuous* on a set $S \subseteq \mathbb{R}$ if, for each $\varepsilon > 0$ and all $x, x' \in S$, there is a $\delta > 0$ such that

$$|x - x'| < \delta \Rightarrow |f(x) - f(x')| < \varepsilon.$$

On an open interval S a continuous function may very well fail to be uniformly continuous. For example, $f(x) = 1/x$ is continuous but not uniformly continuous on (0,1), because the difference $\frac{1}{x} - \frac{1}{x+\delta}$ grows beyond all bounds as x approaches 0. (So this example also shows that the extreme value theorem fails on open intervals, as is obvious from the graph in figure 3.1. Related to this, the Heine-Borel theorem can also fail on open intervals.)

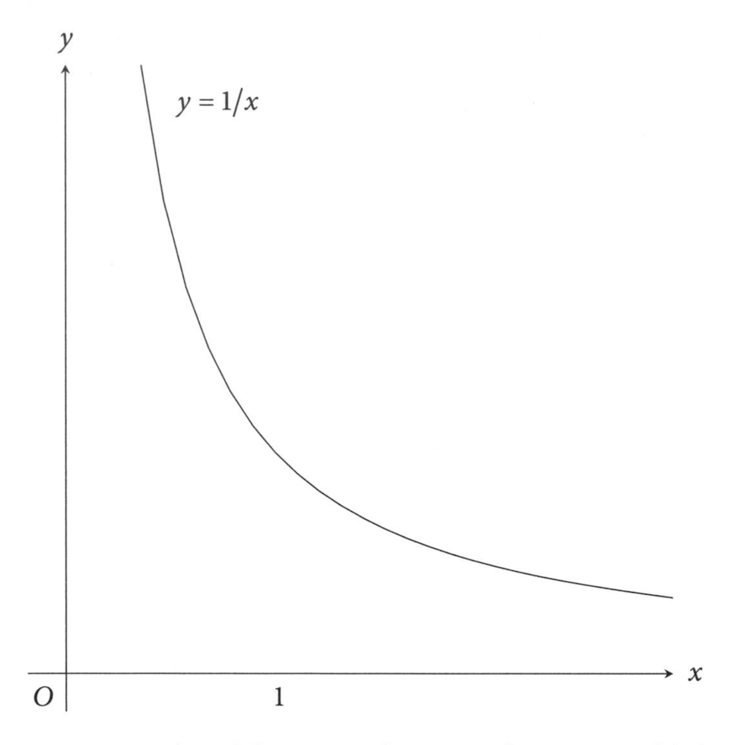

Figure 3.1 : Unbounded continuous function on the open interval (0,1)

However, a continuous function on a *closed* interval is uniformly continuous. It suffices to prove this for the interval [0,1].

Uniform continuity on closed intervals. *If f is continuous on $[0,1]$ then f is uniformly continuous there.*

Proof. Given $\varepsilon > 0$ and any $c \in [0,1]$, continuity of f gives a $\delta(c) > 0$ such that

$$x, x' \in (c - \delta(c), c + \delta(c)) \Rightarrow |f(x) - f(x')| < \varepsilon.$$

The open intervals $(c - \delta(c), c + \delta(c))$, for all $c \in [0,1]$, cover [0,1]. So, by the Heine-Borel theorem, some finite collection of them also covers [0,1]. Call the finitely many covering intervals $I_1, I_2, \ldots, I_n$.

Then if x, x' lie in the same I_k we have $|f(x) - f(x')| < \varepsilon$.

Since $I_1, I_2, \ldots, I_n$ are open intervals, any two of them that overlap have an open interval in common. We let δ be the *minimum length* of the overlaps among $I_1, I_2, \ldots, I_n$.

Now if x, x' are not in the same I_k there is at least one interval of overlap lying between them, and hence $|x - x'| \geq \delta$. Consequently,

$$|x - x'| < \delta \Rightarrow x, x' \in \text{same } I_k \Rightarrow |f(x) - f(x')| < \varepsilon,$$

which shows that f is uniformly continuous. □

Remark. It is worth seeing how this proof can be modified so as to appeal only to the *sequential* Heine-Borel theorem mentioned in section 3.5. Namely, inside each interval $(c - \delta(c), c + \delta(c))$ we can choose rational numbers c^*, d^* with

$$c - \delta(c) < c^* < c < d^* < c + \delta(c),$$

since any real number has rational numbers arbitrarily close to it. Then the rational intervals (c^*, d^*) cover [0,1], because they cover each $c \in [0,1]$. Also, since $(c^*, d^*) \subset (c - \delta(c), c + \delta(c))$ we have

$$x, x' \in (c^*, d^*) \Rightarrow |f(x) - f(x')| < \varepsilon,$$

so we can argue as before that f is uniformly continuous.

But now, the intervals (c^*, d^*) can be put in a sequence, since they correspond to pairs of rational numbers, and hence (by the encoding of pairs of natural numbers by natural numbers explained in section 2.4) to natural numbers. Thus it now suffices to appeal to the sequential Heine-Borel theorem (and so we get around the difficulty that arbitrary sets of intervals cannot be arithmetized).

Riemann Integrability

It follows from the above proof that if f is continuous on [0,1], and any $\varepsilon > 0$ is given, then we can divide [0,1] at points $0 = c_0 < c_1 < c_2 < \cdots < c_{m+1} = 1$ in such a way that

$$c_i \leq x, y \leq c_{i+1} \Rightarrow |f(x) - f(y)| < \varepsilon.$$

We can therefore fit the graph of $y = f(x)$ between the graphs of the *step functions* whose values differ by at most ε for all x in [0,1] (see figure 3.2). The lower step function has the constant value, on $[c_i, c_{i+1})$, equal to the minimum value of f on $[c_i, c_{i+1}]$ (which exists by the extreme value theorem). The upper step function has the constant value, on $[c_i, c_{i+1})$, equal to the maximum value of f on $[c_i, c_{i+1}]$.

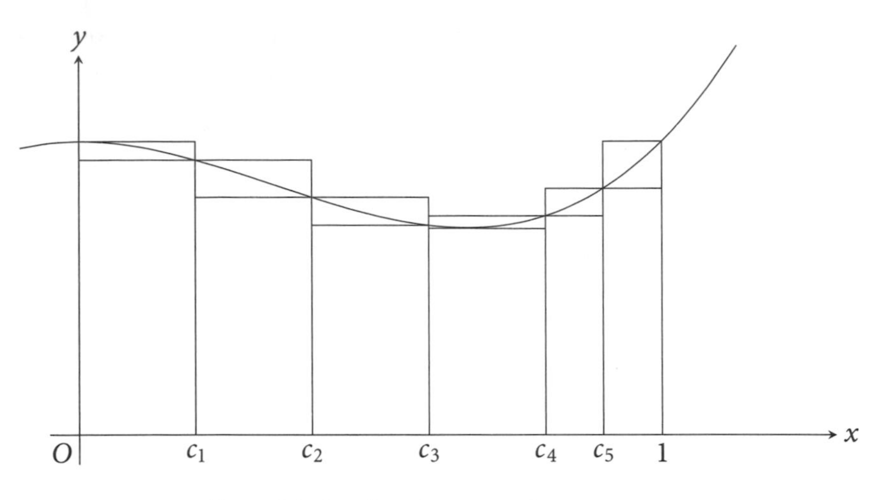

Figure 3.2 : Step functions approximating a curve

The area under each step function is well-defined, being just a union of finitely many rectangles, with the values of the upper areas ≥ the values of all the lower areas. Also, the difference between these areas can be made ≤ any given ε, as we have just seen, so there is a unique value that lies between them, called the *Riemann integral of* f on [0,1], $\int_0^1 f(x)\,dx$.

Thus uniform continuity of continuous functions on [0,1] has the corollary that *each continuous function on a closed interval is Riemann integrable.*

3.8 THE CANTOR SET

An important construction involving sequences of nested intervals is the so-called *Cantor set*, or *middle third set*.[1] It is constructed by removing the open middle third, $(1/3, 2/3)$ of $[0, 1]$, then removing the open middle thirds of the closed intervals that remain, and so on indefinitely. The points of the Cantor set are those common to all the sets of closed intervals occurring in this infinite construction. The sets of intervals obtained in the first six stages are shown in figure 3.3.

The points of the Cantor set correspond to infinite paths in the tree shown in figure 3.4, called a complete binary tree.

[1]The idea, though not precisely the same construction, first appears in Smith (1875).

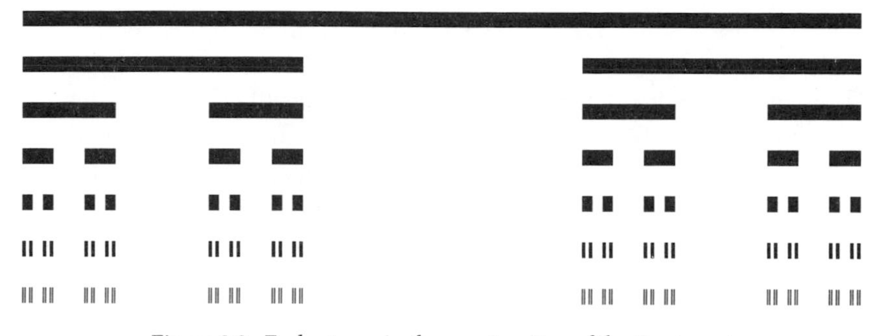

Figure 3.3 : Early stages in the construction of the Cantor set

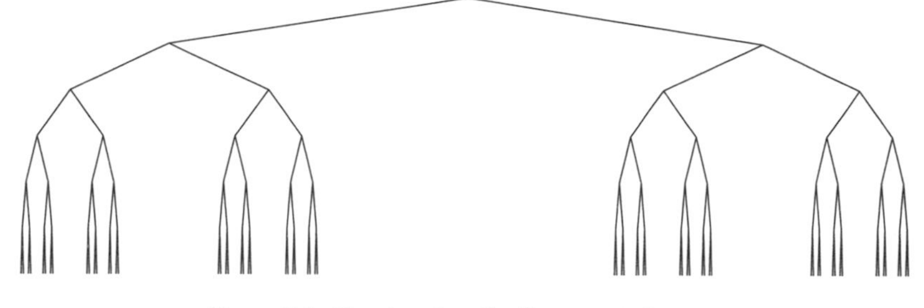

Figure 3.4 : Constructing the Cantor set via a tree

We could label the vertices of this tree by finite sequences of 0s and 1s (0 for "left" and 1 for "right"), so that each infinite path corresponds to an infinite sequence of 0s and 1s. However, it is cleverer to replace each 1 by a 2, because then each infinite path is given by the *ternary* (base 3) *expansion* of the corresponding real number. This is because the left third of [0,1] consists of numbers whose ternary expansion begins with 0, the right third consists of those whose ternary expansion begins with 2, and the left and right thirds of a subinterval give numbers whose next ternary digit is 0 or 2 respectively. Thus *the points of the Cantor set are exactly those with ternary expansions containing only the digits* 0 *and* 2.

Examples are the rightmost path in the left half of the tree, described by the ternary expansion $\cdot 022222\cdots$, which represents the point 1/3, and the leftmost path in the right half of the tree, described by the ternary expansion $\cdot 200000\cdots$, which represents the point 2/3. More generally, any point in the Cantor set corresponds to an infinite nested sequence of intervals, to an infinite path in the binary tree, and to an infinite ternary expansion containing only the digits 0 and 2.

3.9 TREES IN ANALYSIS

The reader will now have noticed that several proofs in this chapter depend on repeated bisection of closed intervals, and the subsequent determination of points by nested sequences of intervals whose lengths tend to zero. The set of all intervals obtainable by repeated bisection can be conveniently viewed as a *tree*, as shown in figure 3.5.

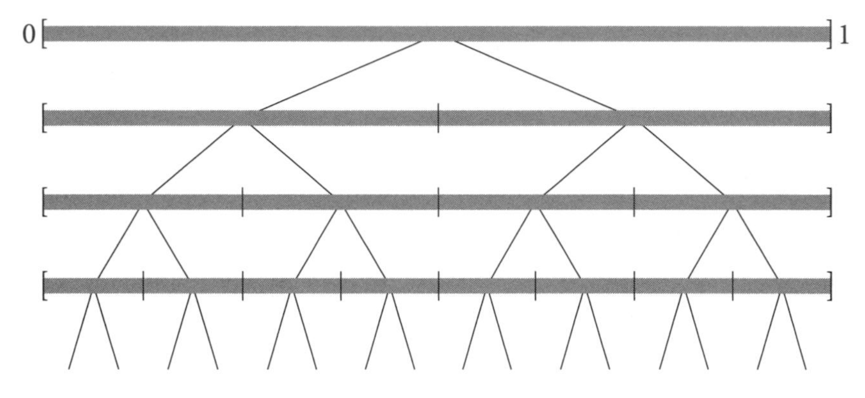

Figure 3.5 : The complete binary tree of bisected intervals

This tree has the whole interval (typically [0,1]) as its top vertex, and the vertices below it are the subintervals obtained by bisection. Thus each vertex has two vertices below it, which is why the tree is called the *complete binary tree B*. Points of [0,1] correspond to infinite nested sequences of subintervals, and hence to *infinite paths* in B. For example, the leftmost infinite path in the tree corresponds to the sequence

$$[0,1] \supset [0,1/2] \supset [0,1/4] \supset [0,1/8] \supset \cdots,$$

which determines the point 0.

In the proofs above we appeal to special arguments to find infinite paths, but there is actually a simple general criterion for their existence, due to Kőnig (1927), which concerns *finitely branching trees*. Such a tree T can be defined as a graph with top vertex v_0, connected by edges to finitely many new vertices $v_1, \ldots, v_k$, and in general with each vertex v_m connected by edges to finitely many new vertices v_n—and these are the only edges. Figure 3.6 shows an example.

The main theorem of Kőnig (1927), which we will call the *Kőnig infinity lemma* or the *strong* Kőnig lemma, states that *if a finitely branching tree*

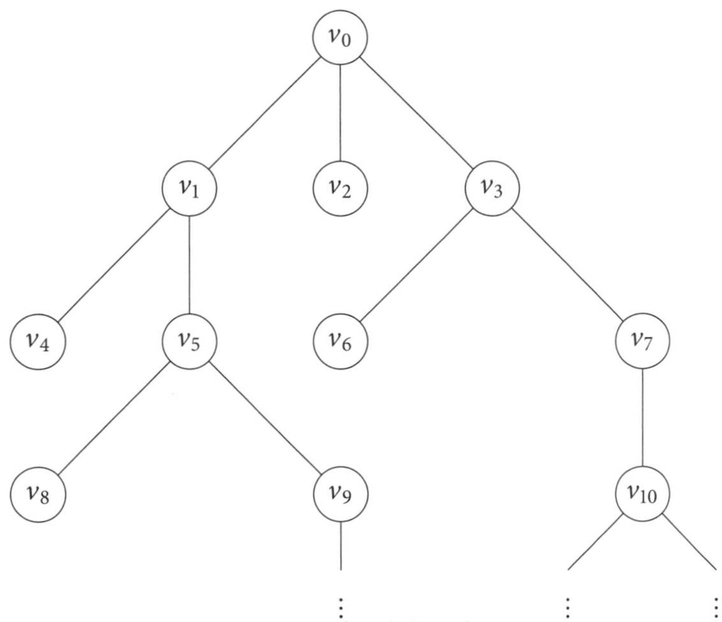

Figure 3.6 : A finitely branching tree

has infinitely many vertices, then it has an infinite path. Its proof, which resembles the argument for Bolzano-Weierstrass or Heine-Borel, involves repeated division of an infinite set into finitely many parts. Namely, since T has infinitely many vertices, one of the finitely many edges out of v_0 leads into a subtree T_1 with infinitely many vertices. By the same argument, one of the finitely many vertices out of the top vertex of T_1 leads into a subtree T_2 of T_1 with infinitely many vertices. Repeating this argument indefinitely, we obtain an infinite path in T.

The *weak Kőnig lemma* is the special case where T is a subtree of the complete binary tree. The results earlier in this chapter suggest that the weak Kőnig lemma is the principle underlying many basic theorems of analysis. This will be confirmed in chapter 7 though, rather surprisingly, the Bolzano-Weierstrass theorem turns out to be equivalent to the strong Kőnig lemma, not the weak one.

The strong Kőnig lemma is also equivalent to several completeness properties of $\mathbb{R}$, such as the least upper bound property and the Cauchy convergence criterion of section 2.3. Thus *trees are a key concept of analysis*—a fact that was not much appreciated until reverse mathematics brought it to light.

Arithmetization of Trees

It is clear that only finitely many vertices of a binary tree, or of a finitely branching tree, are a finite number of edges away from the top vertex. So the vertices can be enumerated by listing those one edge away from the top, then those two edges away, and so on. This means that such trees can be encoded by sets of natural numbers, and hence brought within the scope of the arithmetization project described in the previous chapter.

We will discuss specific methods for arithmetizing trees in section 6.4. For now, it suffices to see how trees can be encoded by sets of *words*, or strings of symbols. The complete binary tree has vertices most naturally encoded by strings of 0s and 1s, as shown in figure 3.7.

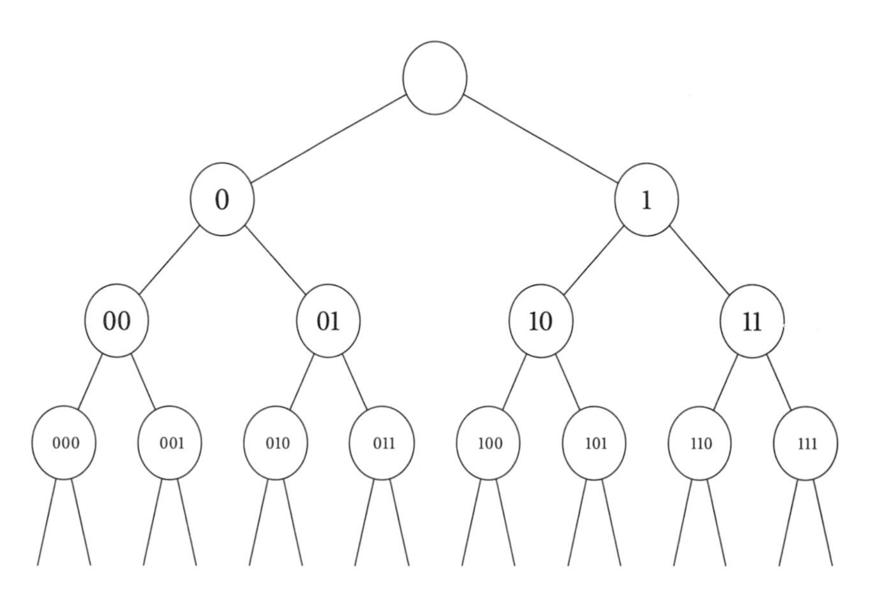

Figure 3.7 : Labeling the vertices of the complete binary tree

The top vertex is labeled by the empty string, those below it by 0 (on the left) and 1 (on the right). In general, the vertices below the vertex labeled σ are labeled $\sigma 0$ (on the left) and $\sigma 1$ (on the right). A *binary tree* T can now be defined as a subset of this set of binary strings with the property that if $\sigma 0 \in T$ or $\sigma 1 \in T$ then $\sigma \in T$. Figure 3.8 on the following page shows an example where $T = \{\text{empty}, 0, 1, 00, 01, 10, 010, 100, 101, \ldots\}$.

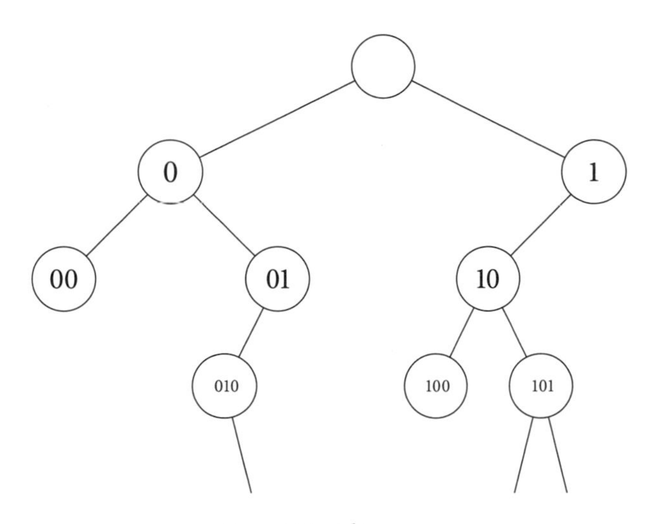

Figure 3.8 : A binary tree

CHAPTER 4

Computability

This chapter foreshadows a constructive approach to analysis, in chapter 7, using a system called RCA_0. The initials RCA stand for "recursive comprehension axiom," and in this context "recursive" means "computable." The goal of RCA_0 is to capture the basic concepts of analysis—real numbers and continuous functions—using *computable* operations on rational numbers. To prepare for RCA_0 we need to study computable sequences and computable sets of rational numbers.

Here we will develop the basic results of computability theory, many of which are about *non*computable sequences and sets, with the goal of revealing the limits of computable analysis. Two of the key examples are a bounded computable sequence of rational numbers whose limit is not computable, and a computable tree with no computable infinite path.

Computability is an unusual mathematical concept, because it is most easily used in an informal way. One often talks about it in terms of human activities, such as making lists, rather than by applying a precise definition. Nevertheless, there *is* a precise definition of computability, so our informal description of computations can be formalized. We describe two such formalizations in the next chapter, and outline a proof of their equivalence.

When it comes to applying the concept of computability in analysis, the most appropriate definition of *computably enumerable set* is one mentioned in section 2.8: that of a Σ_1^0 set. This agreement between the concept of computability and Σ_1^0—the simplest class of arithmetically definable sets—suggests that analysis and computability have a common arithmetical basis, which we will explore further in the next chapter.

4.1 COMPUTABILITY AND CHURCH'S THESIS

Around the year 1900, mathematicians began to pose problems about the existence of *algorithms*. A famous example was Hilbert's tenth problem, so-called because it was tenth on the list of problems that Hilbert presented to the International Congress of Mathematicians in Paris in 1900. An English translation of the problem, which may be found in Hilbert (1902), reads as follows:

> Given a diophantine equation with any number of unknown quantities and with rational integral numerical coefficients: to devise a process according to which it can be determined by a finite number of operations whether the equation is solvable in rational integers.

By "diophantine equation" Hilbert means a polynomial equation for which integer solutions are sought. His "process" determining existence of a solution "by a finite number of operations" is what we would call an *algorithm*. We might also say that Hilbert was calling for a *computer program* to decide existence of integer solutions for arbitrary polynomial equations with integer coefficients.

The discovery of such an algorithm would have been a positive solution to Hilbert's tenth problem, but the solution turned out to be negative: *there is no algorithm that decides, for any arbitrary polynomial* $p(x, y, z, \ldots)$, *whether the equation* $p(x, y, z, \ldots) = 0$ *has a solution in integers* $x, y, z, \ldots$. This result, due to Matijasevič (1971), could only be proved after a mathematical definition of algorithm had been found—something not even known to be possible in 1900. The first definitions of algorithm were discovered by Post in the 1920s, in an analysis of the processes of formal logic. But Post declined to publish them (see Post (1941)) because it seemed impossible to prove that the vague concept of algorithm, or computation, was completely captured by his definitions. It was only after Church (1936b) and Turing (1936) independently proposed definitions—equivalent to each other and to the definitions of Post—that mathematicians became sure that the concept of algorithm really had been captured.

A precise definition of algorithm will be given in the next chapter, along with Turing's decisive analysis of the concept of computation. Here it is more important to understand certain *general characteristics* of any definition, namely:

1. Each algorithm can be written as a finite sequence of symbols in a finite alphabet.
2. Each algorithm can receive *inputs*, which are also finite strings of symbols in a finite alphabet.
3. On each input, the algorithm performs a series of *steps*, always the same steps for the same input. The sequence of steps may be called the *computation* of the algorithm on the given input.
4. If the computation terminates, there is an *output* string, which may be interpreted as a function of the input string.
5. When algorithms are given as strings of symbols they have a uniform interpretation; so there is a *universal algorithm* which, given an algorithm A and input I, will reproduce the computation of algorithm A on input I.

Two of the most important types of algorithms are:

- Where the input and output strings are numerals, in which case the algorithm defines a *computable function* (of positive integers).
- Where the input strings form a set $\mathcal{P}$ of questions with yes/no answers, and the output strings are the words YES and NO. In this case we have an algorithm for the *problem* $\mathcal{P}$. The algorithm is said to *solve* $\mathcal{P}$ if it outputs the correct answer to each question in $\mathcal{P}$. We also say that the set of questions with answer YES is a *computable set.*

Notice that we do not always require an algorithm's computation to halt for each input. The reason for this liberal definition is that deciding whether an algorithm halts for each input is itself a serious problem—in fact, *it is a problem that no algorithm can solve*, as we will see shortly. It follows, as we will also see, that any attempt to restrict the class of algorithms to those that halt on all inputs will be incomplete. The only way to capture the concept of algorithm completely is to include algorithms that sometimes do not halt.

It follows that the functions computable by algorithms include some whose domain is only part of $\mathbb{N}$. For that reason, they are called *computable partial functions.* The term *computable function* is usually reserved for those whose domain is $\mathbb{N}$, but we will also call them *computable total functions* when we wish to emphasize that the function is defined for all positive integers.

The assumption that the various definitions of algorithm by Post, Church, and Turing capture the informal concept of algorithm is known as *Church's thesis*, because it was proposed by Church (1936b). As noticed above, this assumption is needed when we wish to prove *nonexistence* of an algorithm. But Church's thesis also enables us to be informal in proving the *existence* of algorithms. If we can describe an algorithm informally, then we can be confident that the algorithm also has a formal description.

An important feature of algorithms, which is not obvious from the informal notion, is that they can be encoded in the language of arithmetic. We will say more about this in chapter 5. For the moment we note only a superficial, but useful, connection between algorithms and numbers: *there is a computable list* $A_1, A_2, A_3, \ldots$ *of all algorithms*, so each algorithm can be given a number. Since each algorithm is represented by a string of letters in a finite alphabet, it suffices to enumerate all the strings. This can be done by listing the one-letter words first (say, in alphabetical order), then the two-letter words, and so on. For all the known definitions of algorithm it is easily decided whether a string of symbols is a meaningful algorithm—this is what a computer does when it checks whether a program is syntactically correct—so we can omit any meaningless strings and thereby compute a list $A_1, A_2, A_3, \ldots$ of all the algorithms.

This much is easy, in principle. The hard part is to look at an algorithm and decide what it actually *computes*, if anything.

4.2 THE HALTING PROBLEM

Consider the problem consisting of the following questions, which we call the *self-examination problem*.

Q_n: Does algorithm A_n, on input n, output the answer NO?

Suppose A is an algorithm that solves this problem. It is fair to assume that A is given the number n in lieu of the question Q_n, since the question can be reconstructed from the number n. Thus, for each input n, A outputs YES if A_n with input n gives output NO, and otherwise A outputs NO.

Now, since A is an algorithm, we have $A = A_m$ for some number m. What can A_m do on input m? If A_m outputs NO then the answer to the question Q_m is yes, so A_m must output YES, which is a contradiction. There is a similar contradiction if the answer to Q_m is no, hence the algorithm $A = A_m$ cannot correctly answer question Q_m. Thus A fails to solve the self-examination problem, and we have a contradiction.

I call this problem the "self-examination problem," because we want to know what happens when an algorithm is applied to its own number, which is effectively the algorithm itself. If A_n comes to a halt, then we can see what the output is. So *the only thing preventing us from knowing the answer to question* Q_n *is knowing whether* A_n *halts* for a given input. The latter is called the *halting problem*, and it must be unsolvable because otherwise we could solve the self-examination problem.

The halting problem was first proved unsolvable by Turing (1936) (with Turing's own definition of computation, but by a rather similar argument). As we will see, the "self-reference" idea behind the unsolvability proof is the key to many other proofs of noncomputability or unsolvability.

4.3 COMPUTABLY ENUMERABLE SETS

Closely related to the concepts of computable function and solvable problem is the concept of *computably enumerable set*. Informally, a set X is called computably enumerable if there is a computation (typically, a nonhalting computation) which produces a *list* $x_1, x_2, x_3, \ldots$ of all the members of X. There are several equivalent ways of defining a computably enumerable set X in terms of the concept of computable function.

1. X (if it is not empty) is the range of a computable total function f whose domain is the positive integers. In this case we obtain a list of members of X as the list $f(1), f(2), f(3), \ldots$.
2. If infinite, X is the range of an *injective* (that is one-to-one) computable total function g whose domain is the positive integers. Namely, compute $f(1), f(2), f(3), \ldots$ as in the previous definition, but do not put $f(n)$ on the list until it has been checked that $f(n)$ differs from all values of f previously listed. Let $g(m)$ be the mth number put on the list.
3. X is the *domain* of a computable partial function Φ, where Φ is computed by the following algorithm. Given input k, compute the values $f(1), f(2), f(3), \ldots$ in succession. If one of these is found to be k, let $\Phi(k) = 1$.

Conversely, if Φ is any computable partial function, the members of its domain can be listed by a computation in "stages" as follows. At stage n, do n steps in the computations of each of $\Phi(1), \ldots, \Phi(n)$. If the computation of any $\Phi(k)$ halts at this stage, put k on the list.

In the three definitions above the members $x_1, x_2, x_3, \ldots$, or $f(1), f(2), f(3), \ldots$, are not assumed to be positive integers. They could be rational numbers, or any other objects namable by words in a finite alphabet. However, there is no real loss of generality in assuming them to be positive integers, since words in finite alphabet can be encoded by positive integers, by enumerating them in order of length and (for each length) in alphabetical order. From now on we will assume, unless otherwise stated, that members of computably enumerable sets are positive integers.

Among the computably enumerable sets X are those for which the *membership problem* is solvable. That is, there is an algorithm which decides, for each positive integer k, whether $k \in X$. Such a set is called *computable*. Equivalent definitions of computable set in terms of the above concepts are:

1. The *characteristic function* of X, namely

$$x(n) = \begin{cases} 1 & \text{if } n \in X \\ 0 & \text{if } n \notin X, \end{cases}$$

   is computable.
2. If infinite, X is the range of an *increasing* computable function f of positive integers.
   To compute $f(n)$, compute $x(1), x(2), x(3), \ldots$ in turn until n values $x(i)$ have been found equal to 1. If $x(m)$ is the nth value that equals 1, set $f(n) = m$. (Thus m is the nth member of X, in increasing order.) Then the range of f equals X.
   Conversely, if X is the range of an increasing computable function f, we can decide whether a given m belongs to X by computing the values $f(1), f(2), f(3), \ldots$ until a value $f(n) \geq m$ is found. Then $m \in X$ if and only if one of these values equals m.
3. X and its complement $\mathbb{N} - X$ are both computably enumerable.
   If both these sets are computably enumerable, run enumerations of both sets simultaneously. The two sets together include any positive integer, so any given n will eventually appear in one of them. When it does, we will see whether n belongs to X or not.

Conversely, if X is computable, with characteristic function x, we can compute enumerations of both X and $\mathbb{N}-X$. Namely, compute the values $x(1), x(2), x(3), \ldots$, and list n in X if $x(n)=1$, in $\mathbb{N}-X$ if $x(n)=0$.

All of these results appeared in Post (1944), the first paper on computably enumerable sets (then called "recursively enumerable sets," since the word "recursive" was used where we now use "computable"). Post also found examples of computably enumerable sets that are not computable. His basic example was obtained by a "self-reference" argument like that used to prove unsolvability of the "self-examination problem" in the previous section.

To describe Post's basic example, and others we will construct later, we introduce a notation for computable partial functions. Given an enumeration $A_1, A_2, A_3, \ldots$ of algorithms, let Φ_k be the computable partial function of positive integers computed by algorithm A_k. Thus the value of $\Phi_k(n)$ is the output (if any) when algorithm A_k is given the numeral for n as input.

It follows from the universality property 5 of algorithms in section 4.1 that $\Phi_k(n)$ is computable as a function of the two variables k and n. To compute $\Phi_k(n)$ one generates the list of algorithms as far as A_k, then runs A_k on the input n. Now we are ready for Post's example, which gives a computably enumerable set with unsolvable membership problem.

Computably enumerable but noncomputable set. *If* $D=\{k:\Phi_k(k)=0\}$ *then* D *is computably enumerable but not computable.*

Proof. Since $\Phi_k(n)$ is a computable partial function of k and n, $\Phi_k(k)$ is a computable partial function. Thus D is computably enumerable: we list its members in the computation which, at stage n, does n steps in the computations of $\Phi_1(1), \ldots, \Phi_n(n)$ and lists k at any stage when $\Phi_k(k)=0$ is found.

Now suppose that D is computable. Then its characteristic function,

$$d(m)=\begin{cases}1 & \text{if } m\in D\\ 0 & \text{if } m\notin D,\end{cases}$$

is computable, so $d=\Phi_k$ for some k. But then we have the contradiction

$$k\notin D \Leftrightarrow d(k)=0 \Leftrightarrow \Phi_k(k)=0 \Leftrightarrow k\in D,$$

by the definition of D. Hence D is *not* computable. □

The existence of a computably enumerable but noncomputable set influences the treatment of analysis in the system RCA_0, which admits only computable sets. Many naturally arising sequences $r_1, r_2, r_3, \ldots$ (of rational numbers, typically) are computably enumerable but not necessarily computable. To "represent" such a sequence in RCA_0, we encode it by the set of pairs $S = \{\langle n, r_n \rangle : n \in \mathbb{N}\}$. Since $r_1, r_2, r_3, \ldots$ is computably enumerable, there is a computable function f with $f(n) = r_n$. The set S of pairs is then computable, because we can decide whether a pair of the form $\langle n, r \rangle$ is in S by computing $f(n)$ and seeing whether it equals r. Computing the *limit* of the sequence $r_1, r_2, r_3, \ldots$, however, is a different story.

4.4 COMPUTABLE SEQUENCES IN ANALYSIS

Computable objects are the most "concrete" infinite objects, so it would be nice if analysis dealt only with computable real numbers and computable functions. The best-known irrational numbers, such as $\sqrt{2}$, π, and e are in fact computable, in the sense that the nth digit in their decimal expansion is a computable function of n. We will explore "computable analysis" further in chapter 7. However, it is easy to see that analysis cannot be completely computable. If we take a computable sequence of rational numbers $r_1, r_2, r_3, \ldots$ to be one for which r_n is a computable function of n, then we have:

Computable sequence of rationals with a noncomputable limit. *There is a computable sequence of rational numbers with limit whose nth binary digit is not a computable function of n.*

Proof. Take an injective computable total function f with range D, where D is a computably enumerable but noncomputable set (such as the one found in the previous section). This gives a computable sequence $r_1, r_2, r_3, \ldots$ of rational numbers r_n, where

$$r_n = \sum_{i=1}^{n} 2^{-f(i)}.$$

Indeed, r_n has a finite binary expansion with a 1 in the places $f(1), \ldots, f(n)$ and zeros elsewhere. And the binary expansion of the limit has 1 in the kth place for $k \in D$, and zeros elsewhere, so this binary expansion encodes the characteristic function d of D. We know that the characteristic function of D is not computable, so neither is the binary expansion of the limit. □

This example in fact shows more: a computable *increasing* sequence of rational numbers may fail to have a computable least upper bound, because it is obvious from the definition of r_n that $r_{n+1} > r_n$. So the computable real numbers are not "complete" in the classic sense, which goes back to Bolzano (1817). Bolzano assumed that any bounded set of real numbers has a least upper bound in his proof of the intermediate value theorem, and Dedekind (1872)—writing up an idea he had in 1858—defined real numbers in such a way as to make this least upper bound property almost obvious, as we saw in section 2.3. Of course, these results were proved long before computability was understood or even thought to be an issue in analysis.

It is because of the failure of "least upper bound completeness" that we adopted the nested interval concept of completeness in the previous chapter. As we will see in chapter 7, for a computable sequence of closed nested intervals, with a single common point, the common point is computable. Thus, when analysis is limited to computable operations, in the system RCA_0 of chapter 7, the nested interval concept of real number is available.

However, the above example shows that *the least upper bound principle is not provable in* RCA_0. We can encode the computable sequence above by the computable set of pairs $\langle n, r_n \rangle$, which belongs to the model of RCA_0 we construct in section 5.10, whose sets are all the computable sets. But the limit point of the sequence, being noncomputable, does *not* belong to the model, so the model does not satisfy the sentence "every bounded sequence has a least upper bound."

4.5 COMPUTABLE TREE WITH NO COMPUTABLE PATH

In section 3.9 we saw that many basic results of analysis stem from the *weak Kőnig lemma* stating that an infinite binary tree has an infinite path. These results are problematic in computable analysis, and in fact a new axiom is needed to prove them, because a computable infinite tree need not have any computable infinite path. In this section we give an example.

We first construct a pair of computably enumerable sets A, B that are *computably inseparable*. A computable total function f is said to *separate* sets A and B if f takes only the values 0 and 1 and

$$f(n) = \begin{cases} 1 & \text{if } n \in A, \\ 0 & \text{if } n \in B. \end{cases}$$

(Informally, f is a machine that says YES or NO for each input number, always saying YES for numbers in A and NO for numbers in B.) We define A and B using the numbering of computable partial functions $\Phi_1, \Phi_2, \Phi_3, \ldots$ from the previous section; namely

$$A = \{k : \Phi_k(k) = 0\},$$
$$B = \{k : \Phi_k(k) = 1\}.$$

These sets defeat any computable function Φ that tries to separate them. Any computable $f = \Phi_k$ for some k, so if f separates A and B we have

$$\Phi_k(k) = 0 \Rightarrow k \in B \Rightarrow \Phi_k(k) = 1,$$
$$\Phi_k(k) = 1 \Rightarrow k \in A \Rightarrow \Phi_k(k) = 0.$$

Thus either value of $\Phi_k(k)$ is contradictory, and therefore no computable f separates A and B.

We now use the sets A and B to construct an infinite binary tree T, whose infinite paths separate A and B when we interpret paths as sequences of 0s and 1s. The infinite paths are therefore not computable.

Computable tree with no computable infinite path. *There is an infinite binary tree T, the vertices of which form a computable set, but whose infinite paths all separate the sets A and B.*

Proof. We view the complete binary tree as the set of all finite sequences of 0s and 1s, as in section 3.9. A subset T of the vertices is a *subtree* if any vertex above a member of T is also in T. An *infinite path* in T is an infinite sequence of 0s and 1s—that is, a function $\sigma(n)$ whose values are 0 or 1—whose initial segments are all vertices of T. We will construct a tree T whose infinite paths are functions σ that *separate* A and B in the sense that $\sigma(n) = 1$ if $n \in A$ and $\sigma(n) = 0$ if $n \in B$.

To compute T, we construct it in stages, deciding at stage n which vertices at level n belong to T. We also enumerate A and B in stages, using the members found by stage n to decide which vertices at level n to put in T. The idea is to pick vertices that "separate" the members of A and B found so far, which we can assume to be $\leq n$. For example, suppose that by stage 5 we have found 1 and 3 in A and 4 in B. Then any vertex of the form $v = 1 * 10 *$ (where $*$ denotes either 0 or 1) "separates" the parts of A and B found so far, because v has value 1 at places 1 and 3 and value 0 at place 4. We therefore put all four vertices of this form in T.

Notice that when v is put in T then all vertices above v are already in T, because they too "separate" the parts of A and B known at earlier stages. It follows that the set T is a tree. Also, T is computable, because we can decide whether a given vertex v is in T by finding the level n of v and then running n stages of the computation just described.

Now consider a path σ in the complete binary tree that separates A from B; that is, $\sigma(n) = 1$ for each $n \in A$ and $\sigma(n) = 0$ for each $n \in B$. The vertex s at level n on σ separates the members of A and B that are $\leq n$, and hence s is in T. Since this is true for all vertices in σ, the whole infinite path σ is contained in T. Conversely, if τ is an infinite path that does *not* separate A from B, then we have either $\tau(n) = 0$ for some $n \in A$ or $\tau(n) = 1$ for some $n \in B$. Then the vertex t of τ at level n, or one somewhere below it on τ, will *not* be put in T, because at the stage when n is listed in A or B it will be seen that t fails to separate A from B. Consequently, a non-separating infinite path τ is not contained in T.

To sum up, the infinite paths in T are precisely those that separate A from B, and hence (by the computable inseparability of A and B) all the infinite paths of T are noncomputable. □

4.6 COMPUTABILITY AND INCOMPLETENESS

The main theme of this book is the search for the "right axioms" to prove important theorems, but a secondary theme is the *failure* of certain axiom systems to prove certain theorems. After all, if there was an obvious axiom system to prove all theorems then the question of the "right axioms" would hardly arise. The question is unavoidable because axiom systems are *inherently* incomplete. Any consistent axiom system for mathematics will fail to prove certain theorems, so we are bound to need new axioms to prove the missing theorems.

As mentioned in section 1.6, incompleteness is closely related to the concept of computation. We can now refine this claim by defining a *formal system* to be an algorithm that computably enumerates theorems. This covers any axiom system actually in use, and any other theorem-generating system that can reasonably claim to be formal.

It is now easy to see how unprovable sentences arise, via the computably enumerable but not computable set D of section 4.3. Since D is computably enumerable its complement is not, so we cannot computably enumerate all true sentences of the form "$n \notin D$." But a formal system, by definition, computably generates theorems, so there is no formal system $\mathcal{F}$ whose theorems include all true sentences of the form "$n \notin D$" (unless

$\mathcal{F}$ is *unsound*, and it generates some false sentences). Thus, if we interpret the concept of formal system broadly—as a machine for generating theorems—the incompleteness of sound formal systems is an almost obvious consequence of the existence of the set D. Incompleteness in this general form was discovered by Post in the 1920s, and he popularized the argument in his paper Post (1944).

Incompleteness was rediscovered by Gödel (1931) by a more technical argument, but in dramatically stronger form: formal systems for *arithmetic*, such as PA, are incomplete. Gödel's incompleteness can be derived from Post's by *arithmetization of computation*. We give more details in the next chapter. The outcome of arithmetization is that sentences about computably enumerable sets can be translated into sentences in the language of PA. So, instead of finding unprovable sentences of the form "$n \notin D$" we find unprovable sentences about addition and multiplication of natural numbers.

Gödel's theorem reveals that incompleteness exists at quite an elementary level, but alas the unprovable sentences exposed by his construction are not intrinsically interesting. They do not reveal any facts about numbers that number theorists wished to know, though they certainly reveal things about PA that logicians wished to know. For more on this, see section 8.3.

4.7 COMPUTABILITY AND ANALYSIS

Interesting unprovable sentences emerge when we expand PA to a system of analysis by introducing variables for sets of natural numbers. Then, as we saw in chapter 2, it becomes possible to talk about infinite sequences and continuous functions, so we can ask which theorems about them are provable. The answer depends on which *set existence axioms* we adopt.

The simplest reasonable set existence axiom says that computable sets of natural numbers exist. The arithmetization of computation shows that computable sets have a natural description in the language of PA: they are the sets that are both Σ_1^0 and Π_1^0, as defined in section 2.8. Thus this set existence axiom is a natural addition to PA, giving the system RCA_0 of "computable analysis." We can immediately see, however, that if only computable sets are required to exist then the resulting axiom system cannot prove the existence of *non*computable sets such as D. More interestingly, it cannot prove the existence of a limit for every monotonic, bounded, sequence of rationals, as we saw in section 4.4. Thus we have a natural instance of incompleteness: the *monotone convergence theorem*, stating that every bounded monotonic sequence has a limit, is not provable in computable analysis.

We develop the system RCA_0 of computable analysis further in section 5.10 and chapter 7. Despite its obvious incompleteness—or indeed because of it—RCA_0 is a very useful system. It can prove only a few important theorems of analysis, such as the intermediate value theorem and the fundamental theorem of algebra. But, crucially, it is able to prove *equivalences* between theorems that it cannot prove outright. For example, RCA_0 can prove that the monotone convergence theorem is equivalent to the Bolzano-Weierstrass theorem.

This makes RCA_0 an ideal base theory for studying theorems of analysis. Precisely because RCA_0 is *not* able to prove certain theorems it is able to *compare their strengths* by finding set existence axioms to which they are equivalent. The monotone convergence theorem, for example, is equivalent to a set existence axiom (called arithmetical comprehension) asserting the existence of all sets definable in the language of PA. We study the main set existence axioms, and their equivalents in analysis, in chapters 6 and 7.

Constructive Approaches to Analysis

In this chapter we have sketched the development of computability theory from its origins in logic to its role in the foundations of analysis. However, computability was an issue in analysis some time before the concept had a precise definition. In fact, it could be said that both modern logic *and* the associated concept of computation arose from nineteenth-century concerns about the foundations of analysis. The key figure in this development was David Hilbert, and a detailed account of his foundational work (the "Hilbert program") may be found in Sieg (2013). Here we give just a brief summary of some key events, which are the following.

1. Criticism of "nonconstructive" mathematics by Kronecker. In the 1880s, Kronecker raised objections to concepts like arbitrary real numbers, which in his opinion were meaningless except in special cases where the number could be given (essentially) by a computation.

2. Kronecker's objections were provoked, in part, by the results about real numbers proposed by Dedekind (defining real numbers noncomputably) and Cantor (showing that there are uncountably many real numbers).
3. Cantor's diagonal argument, when applied to the "set of all sets," led to the paradox of a "set" *larger than* the "set of all sets." This discovery made it clear that the set concept needed clarification if it was to be used as a basis for analysis.
4. Hilbert proposed a program for securing the foundations of mathematics by means of *axiomatics*. He proposed that analysis, in particular, should be developed in an axiom system for the real numbers, and that this system be proved *consistent* by constructive methods that Kronecker would accept. This was essentially *Hilbert's second problem* of Hilbert (1902).
5. In the following decades Hilbert refined his program by describing the problem as one about computation with finite objects, namely, strings of symbols representing mathematical statements. The problem was to show that the process of generating theorems in an axiom system for analysis—by mechanically applying the rules of logic—did not generate the sentence "0=1."
6. Thus the Hilbert program was reduced to a *problem* that Kronecker would have considered meaningful. Indeed, by 1930 it was known that the question of consistency of an axiom system was equivalent (via arithmetization of computation) to a question of elementary number theory (PA).
7. Alas, this reduction of the consistency problem to a question in PA was made by Gödel (1931), who at the same time showed that the question could not be answered in PA! In fact, the consistency of PA itself (a system much weaker than analysis) can be expressed by a sentence Con(PA) in the language of PA—but Con(PA) is *not* provable in PA. This result is called *Gödel's second incompleteness theorem*, and it applies as well to any system that contains PA—such a system cannot prove its own consistency.
8. This theorem of Gödel (1931) derailed the Hilbert program (as well as having an immense impact on logic and computability theory that we discuss further in chapter 8). But, in the meantime, other mathematicians had followed Kronecker in doing analysis as far as possible by constructive methods. One of the most eminent among

> them was Hermann Weyl, whose book (Weyl (1918)) developed most of basic analysis in a system rather like RCA_0 plus the arithmetical comprehension axiom mentioned above.

Of course, Weyl's work predates the definition of computation, so he has intuitively given "constructions" rather than formal computations. But with hindsight we can see that Weyl's constructions are indeed computations, so his book is a precursor of the system we now call ACA_0 (RCA_0 plus the arithmetic comprehension axiom). In chapter 7 we confirm that ACA_0 does indeed prove the standard theorems of basic analysis.

CHAPTER 5

Arithmetization of Computation

In section 2.8 we saw that the Σ_1^0 formulas of Peano arithmetic (PA) define sets that are "computably enumerable" in an intuitive sense. In chapter 4 we studied the intuitive idea of computable enumerability, assuming only that it has *some* formalization where each computably enumerable set may be recovered from a finite "description," and the descriptions themselves are computably enumerable.

Under this assumption, we discovered that *non*computable objects exist: a computably enumerable but noncomputable set, disjoint computably enumerable sets that are computably inseparable, and an infinite computable tree with no computable infinite path. These results reveal the *absence* of certain important objects in computable analysis, such as least upper bounds of some bounded increasing sequences of rational numbers.

Now it is time to explain why Σ_1^0 formulas of PA capture *all* computably enumerable sets, as claimed by Church's thesis of section 4.1. This allows us to capture "computable analysis" in the language of PA, since computable sets and functions are definable in terms of computable enumerability, as we saw in section 4.3.

To justify the claim that Σ_1^0 = "computably enumerable," in this chapter we make a thorough analysis of the concept of computation. We take a precise, but intuitively natural, concept of computation and translate it into the language of PA. The translation is indeed Σ_1^0, but with a slightly different (though equivalent) definition of Σ_1^0.

5.1 FORMAL SYSTEMS

Computation has been part of mathematics for thousands of years, but *computability* was not thought to be a mathematical concept until the twentieth century. It emerged from *formal systems* of mathematics about 100 years ago, particularly the *Principia Mathematica* of Whitehead and Russell (1910). Their aim was to produce completely rigorous proofs by avoiding human errors, such as unconscious assumptions and gaps in reasoning.

To avoid errors, formal proofs proceed from *axioms* by *rules of inference*. Axioms are viewed as strings of symbols, and the rules produce new strings (including the "theorems") from old in a completely mechanical way. This allows correctness of a proof to be verified without knowing what its symbols mean. Indeed, proofs could be checked by a machine, except that suitable machines had not been invented when the first formal systems appeared.

The very idea of a general symbol-manipulation machine was implicit in *Principia Mathematica*, because that system was thought to be capable of generating all the theorems of mathematics. In the early 1920s Emil Post made an analysis of the axioms and rules of *Principia*, recasting them as mechanical rules for producing strings of symbols, among which were the theorems. Post then proceeded to simplify his rules, in the hope of finding a mechanical way to test the truth of any mathematical statement.

Before long (in 1921) he realized that this hope was in vain, and that neither *Principia*, nor any other consistent system, could generate all the theorems of mathematics.[1] However, he did suspect (correctly, as we now believe) that he had discovered how to generate all sets of strings of symbols that *can* be mechanically generated. In other words, he had discovered a mathematically precise formulation of the concept of *computability*, at least as it appears in the concept of *computably enumerable set*.

Post's formal systems, which he called *normal* systems, are now mainly of historical interest. Smullyan (1961) introduced more usable systems in the same style. Smullyan's *elementary formal systems* are not well-known, but (to my knowledge) they are simpler and more elegant than any other formalization of the computability concept. It is also easy to use elementary formal systems to simulate the most popular realization of the computability concept, the *Turing machine* of Turing (1936), as we

[1]This was his anticipation of Gödel's incompleteness theorem; see Post (1941).

will see in section 5.4. For this reason I will follow Smullyan's approach in this chapter.

5.2 SMULLYAN'S ELEMENTARY FORMAL SYSTEMS

Smullyan models his systems on classical formal systems. They have axioms and rules of inference and they are written in a language with constants and variables. But unlike the usual formal systems there are no parentheses. The constants $a, b, c, \ldots$ are simply strung together to form *words*, such as $aa, aba, cba, \ldots$. The variables $x, y, z, \ldots$ stand for arbitrary words (which can be empty), and they can be strung together with themselves or with constants to form variable words. For example, axb represents any word that begins with a and ends with b.

In addition, there are upper case symbols $P, Q, R, \ldots$, called *set variables*, that represent sets or properties. We write Pw to mean "$w \in P$" or "w has property P." We note that w could represent an ordered pair, or an ordered triple, and so on, in which case P could be viewed as representing a binary (or ternary, and so on) *relation* such as a *function*. We do this simply by inserting commas between the members of the pair, or triple, and so on. Finally, there is a symbol $\Rightarrow$ that represents implication.

The purpose of an elementary formal system (EFS) is to generate "theorems" of the form Pw, where w is a constant word, and thus to "computably enumerate" the members of a set P of words. To this end, the system has axioms of two kinds:

Axioms of an EFS.

- Pw, for certain constant words w, stating that these w belong to P.
- $Px_1 \Rightarrow Px_2 \Rightarrow \cdots \Rightarrow Px_n$, for certain variable words $x_1, x_2, \ldots, x_n$. This axiom is read "Px_1 implies that Px_2 implies that $\cdots$ Px_{n-1} implies Px_n." It is logically equivalent to

$$(Px_1 \wedge Px_2 \wedge \cdots \wedge Px_{n-1}) \Rightarrow Px_n.$$

An example, which defines the set E of strings of the form $aa \cdots a$ of even positive length, is

$$Eaa$$
$$Ex \Rightarrow Exaa.$$

It is easy to see that we can derive all (and only) theorems of the form $Eaa \cdots a$ with a positive even number of letters a if we use the obvious rules of inference, which in fact cover all EFS:

Rules of inference of an EFS.

- In any axiom, the result of substituting an arbitrary word for each occurrence of a variable is a theorem.
- If U and $U \Rightarrow V$ are theorems, and if U is not itself of the form $X \Rightarrow Y$, then V is a theorem. (This rule is called *modus ponens* after the similar rule in classical logic.)

(The reason for the restriction on modus ponens is the following. If U *is* $X \Rightarrow Y$, with X and Y containing no arrows, then $U \Rightarrow V$ is $X \Rightarrow Y \Rightarrow V$, which means $(X \wedge Y) \Rightarrow V$. But U is *not* the same as $X \wedge Y$, so V does not follow from U and $U \Rightarrow V$. However, if X and Y are theorems, and we also have $X \Rightarrow Y \Rightarrow V$, then we *can* conclude V, as you would hope.)

Examples of Axiom Systems

1. An EFS generating the set P of palindromes (words spelled the same backwards and forwards) on the alphabet $\{a, b\}$.

$$Pa$$
$$Paa$$
$$Pb$$
$$Pbb$$
$$Px \Rightarrow Paxa$$
$$Px \Rightarrow Pbxb$$

These axioms allow us to start with any one- or two-letter palindrome and to expand it to an arbitrary palindrome by repeatedly attaching the same letter at both ends.

2. An EFS generating the set S of words on $\{a, b, c\}$ which involve only the letters a, b.

$$Sa$$
$$Sb$$
$$Sx \Rightarrow Sy \Rightarrow Sxy$$

These axioms generate any word consisting of as and bs by starting with one-letter words and concatenating arbitrary words.

3. An EFS generating the set T of nonempty words on $\{a, b\}$ with equal numbers of as and bs.

$$Tab$$
$$Tba$$
$$Txyz \Rightarrow Txaybz$$
$$Txyz \Rightarrow Txbyaz$$

The first two axioms give the two-letter words with equal numbers of as and bs. The latter two axioms, since x, y, or z may be empty, allow one a and one b to be inserted at the same time anywhere in a word, thus maintaining equal numbers of as and bs.

5.3 NOTATIONS FOR POSITIVE INTEGERS

Elementary formal systems generate "words" rather than "numbers" because their purpose is to produce theorems or formulas, which are generally strings of symbols in an alphabet with more than one symbol. However, numbers themselves can be represented by strings of symbols—namely, *numerals*—and hence we can interpret words over any finite alphabet as numbers by viewing the words as numerals.

The simplest numerals are the words 1, 11, 111, ... on a single-symbol alphabet, where the positive integer n is represented by a string of n ones. These *base one* or *unary* numerals are simple and natural, but in some ways too simple to be convenient. The main inconvenience is that unary numerals for even modest size numbers, say 87 and 88, are hard to tell apart—because they are so long.

The usual base 10 numerals are strings of symbols from the alphabet $\{0, 1, 2, 3, 4, 5, 6, 7, 8, 9\}$, but they have the disadvantage that many different strings represent the same number. For example, 1, 01, 001, and so on all represent the number 1. The usual base 2 numerals have the same disadvantage. If we are content to have numerals only for *positive* numbers then a solution to this problem is given by what Smullyan calls the *dyadic* system of numerals. This system gives a one-to-one correspondence between positive integers n and strings of the digits 1 and 2. The dyadic numerals for the first few positive integers are the following (with base 10 numerals on the left):

$$
\begin{aligned}
1 &= 1 \\
2 &= 2 \\
3 &= 11 \\
4 &= 12 \\
5 &= 21 \\
6 &= 22 \\
7 &= 111 \\
8 &= 112 \\
9 &= 121 \\
10 &= 122 \\
11 &= 211 \\
12 &= 212 \\
13 &= 221 \\
14 &= 222 \\
&\dots
\end{aligned}
$$

In general, each positive integer n is represented by a string $d_k \cdots d_2 d_1$, where $d_1, d_2, \dots, d_k$ are the unique digits in $\{1, 2\}$ such that

$$n = d_k \cdot 2^{k-1} + \cdots + d_2 \cdot 2 + d_1 \cdot 1.$$

Obviously, each string of 1s and 2s represents a positive integer. Also, we can show inductively that the string for each positive n is unique. Namely, given the unique dyadic numeral for n, the obvious process for adding 1 gives the unique dyadic numeral for $n + 1$. (See section 5.8 for an EFS that does this.)

Symbol strings in any alphabet $\{a_1, a_2, \dots, a_n\}$ can be encoded by dyadic numerals if we replace a_1 by 12, a_2 by 122, a_3 by 1222, and so on. So all the operations on strings that occur in an EFS can be viewed as operations on numbers. This gives a (distant) glimpse of how to *arithmetize computation.*

But before doing so we will study the opposite problem: doing arithmetic in elementary formal systems. This will improve our understanding of the "computational ability" of elementary formal systems, and thus bolster the claim that they can represent any computation. At the same time, we will become more familiar with the computational content of arithmetic, which will be useful when we come to arithmetize computation later.

Universal Elementary Formal Systems

An important side effect of the encoding of a_1 by 12, a_2 by 122, a_3 by 1222, and so on, is that *words in an arbitrary finite alphabet may be encoded by words in the fixed alphabet* $\{1, 2\}$. This raises the possibility of a *universal* EFS that can simulate the operations of any particular EFS. By encoding the alphabet of each EFS in the alphabet $\{1, 2\}$, and perhaps adding a few more letters for convenience, it should be possible to construct an EFS U that is universal in the sense that U generates all true statements of the form "system S generates theorem T," with some appropriate encoding of systems and theorems.

In the Turing machine model of computation the corresponding universal system is called a *universal Turing machine*, and one was constructed in the groundbreaking paper of Turing (1936). In the EFS model of computation, a universal EFS is described in Smullyan (1961), pp. 12–14 and 16–18. We skip the details of a universal EFS, but mention that it may be used to formalize the argument of section 4.3, where we described a computably enumerable set of natural numbers whose complement is not computably enumerable.

5.4 TURING'S ANALYSIS OF COMPUTATION

Before we explore EFS computation in detail, we should look at the classic introduction of computation in Turing (1936). Turing arrived at his concept of computation by analyzing the way that humans compute with pencil and paper. His assumptions (and his reasons for them) about the way a human "computer" operates (or could operate) were the following:

1. The computer can recognize finitely many different *symbols* S_j, and scans one of them at a time. A "symbol" in this sense may be a finite block of digits or letters of the alphabet, but only finitely many such symbols can be distinguished. "Symbols" that are too similar will be confused, for example, 9999999999999999 and 999999999999999.
2. The computer has finitely many *internal states* (think of them as "mental states") q_i for the same reason: if there are infinitely many, then some will be too similar to distinguish.

3. Computation is directed by a finite *program*, which tells what to do when a given symbol is scanned while in a given internal state.
4. It can be assumed that symbols are written on a tape divided into squares, one symbol per square, and that at each *step* of computation a *read/write head* replaces the scanned symbol by another symbol, moves one square to the left or right, and enters another internal state.

The last condition, of course, restricts the way we normally use pencil and paper, but it is a restriction we can endure at the cost of slower computation. For example, we can compute 77489+45132 on a single line by moving back and forth between the two numerals, crossing off digits (that is, replacing a digit symbol such as 3 by the crossed digit symbol 3̸) as their sums are computed and "carrying" digits mentally (that is, by internal states).

With some practice, it becomes clear that all our familiar computations can be done under Turing's conditions, and are "programmable" by Turing machine. The program of a machine can be written as a finite sequence of *quintuples*, of the form $q_iS_jS_kRq_l$ (or $q_iS_jS_kLq_l$). The command $q_iS_jS_kRq_l$ says that if the state is q_i and the scanned symbol is S_j, then replace S_j by S_k, move one square to the right, and go into state q_l. (And similarly if L occurs instead of R.)

It is noteworthy that Post (1936) arrived at virtually the same concept of computation independently, as Turing himself said in Turing (1950). Further details may be found in the papers of Turing and Post, or indeed in many books on the theory of computation. The Turing machine concept was decisive in convincing logicians, notably Gödel, that the intuitive concept of computation could be formalized.

From Turing Machines to Elementary Formal Systems

Another historic paper[2] worth reading is Post (1947), in which the Turing machine concept of computation is translated into one based on *word replacement*. Post shows that each computation by a Turing machine T can be encoded by a sequence of words. The kth word w_k encodes the *machine configuration*, consisting of marked portion of T's tape at step

[2]This paper is famous because it is the first in which unsolvability is proved for a problem posed by a mainstream mathematician; namely, what we now call the *word problem for semigroups*. The word problem was posed by Thue (1914).

k, together with a symbol q_i, inserted to the left of the scanned symbol S_j, denoting the current internal state. (Figure 5.1 shows an example, with $S_j = 7$ and the head in state q_3.)

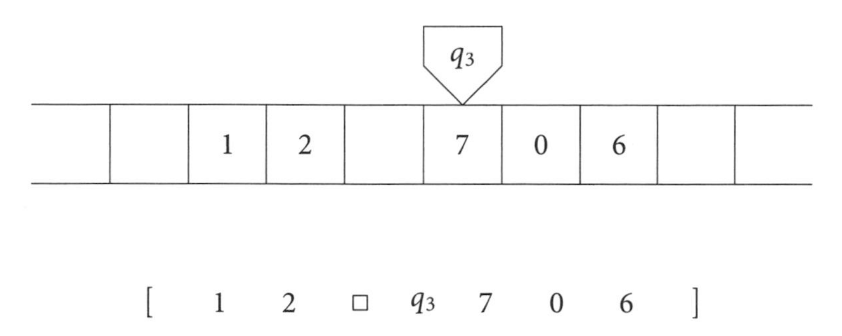

$$[\quad 1 \quad 2 \quad \square \quad q_3 \quad 7 \quad 0 \quad 6 \quad]$$

Figure 5.1 : A machine configuration and the word that encodes it

The word w_{k+1} results from w_k by *replacing* the subword q_iS_j in w_k, together with one or two symbols to its left and right, by a new subword, as required by the program for T. Since there are only finitely many such subwords, only finitely many word transformations are needed to simulate the program for T. For example, if the program for T includes the quintuple $q_3 78Lq_4$ then we need the word replacement rules

$$S_k q_3 7 \to q_4 S_k 8 \quad \text{for each symbol } S_k.$$

In particular, the rule $\square q_3 7 \to q_4 \square 8$ allows the word in figure 5.1,

$$w_k = [12 \square q_3 706], \quad \text{to be replaced by the word} \quad w_{k+1} = [12q_4 \square 806],$$

which encodes the next configuration of the machine.

With a finite set of rules for word replacement we are very close to an elementary formal system. Indeed, a rule $u \to v$ saying that word u may be replaced by word v is implemented in an EFS by the axiom $Wxuy \Rightarrow Wxvy$. This, then, is how Turing machines can be simulated by elementary formal systems. Now it is time to explore EFS computation directly.

5.5 OPERATIONS ON EFS-GENERATED SETS

The intuitive property of computable enumerability is preserved by some of the basic operations on sets, such as union, intersection, and cartesian product. If we suppose that we can generate lists of members of sets S and T, then we can also generate lists of members of:

- $S \cup T$
  namely, by listing S and T simultaneously and merging the two lists.
- $S \cap T$
  namely, by listing S and T simultaneously and making a third list—of the elements that appear in both S and T.
- $S \times T$
  namely, by listing $\langle x, y \rangle$ for each x that appears on the list for S and each y that appears on the list for T.

In terms of *properties*, rather than sets, we are saying that if the properties $S(x)$ and $T(x)$ (corresponding to the properties $x \in S$ and $x \in T$) are computably enumerable, then so are the properties $S(x) \vee T(x)$, $S(x) \wedge T(x)$, and $S(x) \wedge T(y)$.

Definition. A set S (of words in some finite alphabet) is called EFS-*generated* if there is an EFS that proves Sx if and only if $x \in S$.

With this definition we can confirm that the above operations—which preserve computable enumerability in the intuitive sense—also preserve EFS-generated sets.

Operations on EFS-generated sets. *If S and T are EFS-generated sets, then so are $S \cup T$, $S \cap T$, and $S \times T$.*

Proof. Given an EFS to generate S and an EFS to generate T, we construct EFS to generate each of (i) $S \cup T$, (ii) $S \cap T$, and (iii) $S \times T$.

(i) Suppose we have an EFS for each of S and T. By rewriting the system for T, if necessary, we can ensure that the systems have no set variables in common. The union of the two systems will then function as two independent systems, proving the theorems Sx for $x \in S$ and the theorems Tx for $x \in T$.

So if X is a set variable not already used, adding the axioms

$$Sx \Rightarrow Xx, \quad Tx \Rightarrow Xx$$

gives an EFS that proves the theorems Xx for precisely the x in $S \cup T$.

(ii) Similarly, adding the axiom $Sx \Rightarrow Tx \Rightarrow Xx$ gives a system that proves the theorems Xx for exactly the x in $S \cap T$, because $Sx \Rightarrow Tx \Rightarrow Xx$ means $(Sx \wedge Tx) \Rightarrow Xx$.

(iii) To set up a system to generate $S \times T$ we assume, without loss of generality, that the comma is not one of the symbols in the systems for S and T. (If it is, replace it by some other symbol not already used.) This enables us to use the string x, y to represent the ordered pair $\langle x, y\rangle$.

Under this assumption we combine the systems for S and T as above and add the axiom

$$Sx \Rightarrow Ty \Rightarrow Px, y \quad \text{where } P \text{ is a set variable not already used.}$$

Since this axiom means $(Sx \wedge Ty) \Rightarrow Px, y$, the new system proves Px, y for precisely the $\langle x, y\rangle$ in $S \times T$. □

Using the comma as in part (iii) of the proof above, we can speak of EFS-generated sets of n-tuples.

Definition. A set S of ordered n-tuples $\langle x_1, x_2, \ldots, x_n\rangle$ is said to be EFS-generated if there is an EFS whose theorems of the form $Sx_1, x_2, \ldots, x_n$, for comma-free words $x_1, x_2, \ldots, x_n$, are those for which $\langle x_1, x_2, \ldots, x_n\rangle \in S$.

We can now contemplate operations on EFS-generated sets of n-tuples. The most important of these is called *existential quantification* (when speaking in terms of properties) or *projection* (when speaking in terms of sets).

Definition. If $W(x_1, \ldots, x_k, y_1, \ldots, y_l)$ is a property of $(k + l)$-tuples, then the property $\exists x_1 \cdots \exists x_k W(x_1, \ldots, x_k, y_1, \ldots, y_l)$ is an *existential quantification* of the property W, and the set

$$\{\langle y_1, \ldots, y_l\rangle : \exists x_1 \cdots \exists x_k W(x_1, \ldots, x_k, y_1, \ldots, y_l)\}$$

is the corresponding *projection* of the set

$$\{\langle x_1, \ldots, x_k, y_1, \ldots, y_l\rangle : W(x_1, \ldots, x_k, y_1, \ldots, y_l)\}.$$

Projection of EFS-generated sets. *If W is an EFS-generated set of $(k + l)$-tuples, then any projection of W is EFS-generated.*

Proof. Given an EFS for W, let E be a new set variable and add the axiom

$$Wx_1, \cdots, x_k, y_1, \cdots, y_l, \Rightarrow Ey_1, \cdots, y_l. \quad □$$

5.6 GENERATING Σ_1^0 SETS

In section 2.8 we classified the sets definable in Peano arithmetic (PA) by means of the formulas defining them. We began with equations between terms involving variables, the constant 0, and the S, $+$, and $\cdot$ functions, then combined equations by the Boolean operations $\wedge$, $\vee$, and $\neg$ to form *quantifier-free* formulas.

In the present section we follow a similar path, but we enlarge the class of quantifier-free formulas by allowing the *bounded quantifiers* $(\forall x < y)$ and $(\exists x < y)$ to occur in them. The properties defined by the new formulas do not in fact include any not defined by the old, but it is useful to have the extra flexibility afforded by bounded quantifiers.

In particular, if we define Σ_1^0 formulas to be those obtained by existential quantification of the new quantifier-free formulas it becomes easier to reach our ultimate goal of showing that the EFS-generated sets equal the Σ_1^0 sets. The more flexible definition of Σ_1^0 is in fact the one used in the definitive treatment of reverse mathematics, Simpson (2009). In this section and the next we prove one direction of the equality: Σ_1^0 sets are EFS-generated.

Since we have already shown, in the last section, that the existential quantification of an EFS-generated set is EFS-generated, it suffices to show that the sets defined by quantifier-free formulas are EFS-generated. We begin with sets defined by equations, and the related problem of representing *functions* (and relations) by sets—particularly the S, $+$, and $\cdot$ functions.

Definition. A relation $R(x_1, x_2, \ldots, x_n)$ is EFS-*representable* if there is an EFS for the set $\{\langle x_1, x_2, \ldots, x_n\rangle : R(x_1, x_2, \ldots, x_n)\}$.

EFS-representation of basic relations. *The relations*
(i) $S(x) = y$, (ii) $x + y = z$, (iii) $x \cdot y = z$, (iv) $x < y$, (v) $x \leq y$, and (vi) $x \neq y$
are EFS-representable.

Proof. (i) The relation $S(x) = y$ in dyadic numerals is represented by Sx, y in the EFS:

$$S1, 2$$

$$S2, 11$$

$$Sx1, x2$$
$$Sx, y \Rightarrow Sx2, y1$$

These axioms are clearly true when Sx, y is interpreted as $S(x) = y$ and x, y are dyadic numerals. So the instances of Sx, y occuring as theorems of this EFS express true instances of $S(x) = y$. To see why *all* instances occur as theorems we show that all dyadic numerals x occur in theorems Sx, y. The shortest instances $x = 1$ and $x = 2$ are in the first two axioms; the third axiom allows us to attach a 1 to the right of any instance of x; and the fourth axiom (about "carrying 1") allows us to attach a 2 to the right of any instance of x.

(ii) Given the EFS for $S(x) = y$, we obtain an EFS for $x + y = z$ by adding axioms that implement the inductive definition of + (for positive integers). Writing Ax, y, z to represent $x + y = z$, suitable axioms are

$$Sx, u \Rightarrow Ax, 1, u \qquad \text{(base step)}$$

and (using the "and" symbol $\wedge$ for the moment)

$$(Ax, v, w \wedge Sv, y \wedge Sw, z) \Rightarrow Ax, y, z. \qquad \text{(induction step)}$$

The induction step is written officially as the axiom

$$Ax, v, w \Rightarrow Sv, y \Rightarrow Sw, z \Rightarrow Ax, y, z.$$

(iii) Given the above EFS for S and +, we obtain an EFS for $\cdot$ by adding axioms that define $\cdot$ inductively, writing Mx, y, z to represent $x \cdot y = z$:

$$Mx, 1, x \qquad \text{(base step)}$$

and

$$(Mx, v, w \wedge Sv, y \wedge Aw, x, z) \Rightarrow Mx, y, z. \qquad \text{(induction step)}$$

The induction step is written officially as the axiom

$$Mx, v, w \Rightarrow Sv, y \Rightarrow Aw, x, z \Rightarrow Mx, y, z.$$

(iv) Writing Lx, y to represent $x < y$, we can generate all correct instances of Lx, y by adding the following axioms to those in (i):

$$Sx, y \Rightarrow Lx, y$$
$$Lx, y \Rightarrow Ly, z \Rightarrow Lx, z$$

(v) Writing $L'x, y$ to represent $x \leq y$, we get axioms for $x \leq y$ by adding the following to those in (i) and (iv):

$$Lx, y \Rightarrow L'x, y$$
$$L'x, x$$

(vi) The latter axiom (if we wrote it for a different set variable, E say) defines the *equality* relation $x = y$. The *in*equality relation $x \neq y$ can be represented by Nx, y if we add the following axioms to those in (iv):

$$Lx, y \Rightarrow Nx, y$$
$$Ly, x \Rightarrow Nx, y$$

□

5.7 EFS FOR Σ_1^0 RELATIONS

Boolean Combinations of Equations

We first wish to use the theorem above to prove that an equation

$$t_1(x_1, \ldots, x_k) = t_2(y_1, \ldots, y_l),$$

where t_1 and t_2 are terms built from variables and 0 by the S, $+$, and $\cdot$ functions, is an EFS-representable relation between $x_1, \ldots, x_k, y_1, \ldots, y_l$. To do this we need to prove that the *composite* of EFS-representable functions is itself EFS-representable. An example will suffice to show how this is done. Given the relations $x + y = z$ and $z = u \cdot v$ we wish to represent the relation

$$R(u, v, x, y) \iff x + y = u \cdot v.$$

We already have an EFS that represents $x + y = z$ by Ax, y, z and an EFS that represents $u \cdot v = z$ by Mu, v, z. Then all we have to do is add the axiom

$$Ax, y, z \Rightarrow Mu, v, z \Rightarrow Ru, v, x, y.$$

In general, if $R_1x_1, \ldots, x_k, z$ represents the relation $t_1(x_1, \ldots, x_k) = z$ and $R_2y_1, \ldots, y_l, z$ represents the relation $t_2(y_1, \ldots, y_l) = z$, then we can represent the relation $t_1 = t_2$ by the relation $Qx_1, \ldots, x_k, y_1, \ldots, y_l$ if we add the axiom

$$R_1x_1, \ldots, x_k, z \Rightarrow R_2y_1, \ldots, y_l, z \Rightarrow Qx_1, \ldots, x_k, y_1, \ldots, y_l.$$

We can also represent the relation $t_1 \neq t_2$ by writing down the axioms for Nx, y, then adding

$$R_1 x_1, \ldots, x_k, w \Rightarrow R_2 y_1, \ldots, y_l, z \Rightarrow Nw, z.$$

Thus we can find EFS-representations of both relations $t_1 = t_2$ and $t_1 \neq t_2$.

Combining this with the EFS for union and intersection of EFS-generated sets (from section 5.5), we can obtain an EFS *for any Boolean combination of equations between terms*; that is, any combination of equations by means of the connectives $\wedge$, $\vee$, and $\neg$. This is because the equivalences

$$\neg(\varphi \wedge \psi) \Leftrightarrow (\neg\varphi) \vee (\neg\psi),$$
$$\neg(\varphi \vee \psi) \Leftrightarrow (\neg\varphi) \wedge (\neg\psi),$$

allow $\neg$ signs to be "pushed inward" until they hit equations $t_1 = t_2$ and turn them into inequations $t_1 \neq t_2$. At this stage all other connectives are $\vee$ or $\wedge$, shown to be EFS-representable in section 5.5. Since we can represent $t_1 = t_2$ and $t_1 \neq t_2$, we can obtain an EFS for any Boolean combination of equations, as claimed.

Such Boolean combinations were called *quantifier-free* formulas of PA in section 2.8. But we now wish to expand the concept "quantifier-free" to include the *bounded* quantifiers $(\forall x < y)$ and $(\exists x < y)$.[3] To avoid confusing the two possible meanings of "quantifier-free" we will use the term Σ_0^0 *formula* for one built from equations between terms by applying Boolean operations and bounded quantifiers. We will also show that the result of applying a bounded quantifier to an EFS-representable relation is EFS-representable.

Bounded Quantifiers

First observe that a bounded existential quantifier is no problem because

$$(\exists y < z) R(x_1, \ldots, x_k, y) \Leftrightarrow (\exists y)[R(x_1, \ldots, x_k, y) \wedge y < z].$$

We know from section 5.5 that existential quantification of an EFS-representable relation is EFS-representable, and $R(x_1, \ldots, x_k, y) \wedge y < z$ is representable if R is, because $y < z$ is EFS-representable and so is the intersection of EFS-representable relations.

[3]This includes the quantifiers $(\forall x \leq y)$, which is equivalent to $(\forall x < y + 1)$, and $(\exists x \leq y)$, which is equivalent to $(\exists x < y + 1)$.

Thus it remains to prove:

EFS-representation of bounded universal quantification. *If the relation* $R(x_1, \ldots, x_k, y)$ *is EFS-representable so is* $(\forall y < z)R(x_1, \ldots, x_k, y)$.

Proof. We introduce a set variable $R^<$ such that $R^<(x_1, \ldots, x_k, z)$ is equivalent to $(\forall y < z)R(x_1, \ldots, x_k, y)$. The relation $R^<$ vacuously satisfies $R^<(x_1, \ldots, x_k, 1)$ because there is *no* positive integer $y < 1$. The relation also satisfies

$$[R^<(x_1, \ldots, x_k, z) \wedge R(x_1, \ldots, x_k, z) \wedge w = S(z)] \Rightarrow R^<(x_1, \ldots, x_k, w).$$

Thus, if we write down the axioms for the function S from the previous section, then add the EFS for R and also the axioms

$$R^< x_1, \ldots, x_k, 1$$
$$R^< x_1, \ldots, x_k, z \Rightarrow R x_1, \ldots, x_k, z \Rightarrow Sz, w \Rightarrow R^< x_1, \ldots, x_k, w$$

we obtain an EFS that generates all theorems of the form $R^< x_1, \ldots, x_k, z$. □

We now *redefine* Σ_1^0 relations (as stated in the preview to this chapter), to be the existential quantifications of Σ_0^0 relations of PA and we have:

Corollary. *All* Σ_1^0 *relations are EFS-generated.*

Proof. The Σ_0^0 relations are those obtained from equations $t_1 = t_2$, by Boolean operations and bounded quantifications, where the terms t_1 and t_2 result from variables and 0 by applying the S, $+$, and $\cdot$ functions.

The theorems of this section show that Σ_0^0 relations are EFS-generated, and hence so are the Σ_1^0 relations, by the theorem on existential quantification in section 5.5. □

5.8 ARITHMETIZING ELEMENTARY FORMAL SYSTEMS

The previous sections of this chapter have demonstrated the computational ability of EFS, by showing that they can represent all the Σ_1^0 relations of PA. Indeed, the ingredients of Σ_1^0 relations—the S, $+$, and $\cdot$ functions, equations and their Boolean combinations, bounded quantifiers, and existential quantification—are "simulated" in a way that closely tracks their meaning.

When it comes to demonstrating the converse—that Σ_1^0 relations of PA can "simulate" the workings of EFS—there are some technical difficulties that make the simulation harder to track. Because of this, we will carry out only the most fundamental parts of the simulation in detail, and describe the overall process in broader terms, so that the reader has a better chance of seeing the big picture. (Even so, there are some messy details, which may best be skipped at first reading.)

Words and Numbers

As we said in section 5.3, the first step in arithmetizing elementary formal systems is to encode strings of letters (or "words") by dyadic numerals. These numerals have the potential to encode words in an arbitrary alphabet, since we can encode different letters by the numerals 12, 122, 1222, and so on. The hard part is to reflect natural operations on words by operations on numbers, using only the language of PA with its built-in S, $+$, and $\cdot$ functions. The most fundamental operation on words $\boldsymbol{x}$ and $\boldsymbol{y}$ is to form the word $\boldsymbol{xy}$, the *concatenation* of $\boldsymbol{x}$ and $\boldsymbol{y}$, by writing down $\boldsymbol{x}$ and then $\boldsymbol{y}$ to its right.

If x and y are the numbers whose binary numerals are $\boldsymbol{x}$ and $\boldsymbol{y}$ respectively, we let $x^\frown y$ denote the number whose binary numeral is $\boldsymbol{xy}$. We can define the *numerical concatenation function* $^\frown$ via the following series of definitions. Notice that these definitions use at most bounded quantifiers, hence they are Σ_0^0 as defined in the previous section.

1. First we define the relation "x divides y" by
$$x \text{ div } y \Leftrightarrow (x = 1) \vee (\exists z < y)(x \cdot z = y).$$
2. Next
$$x \text{ is a power of } 2 \Leftrightarrow (\forall y < x)[(y \text{ div } x \wedge 1 < y) \Rightarrow 2 \text{ div } y].$$
3. Then, if $l(x)$ = length of dyadic numeral for x, we have
$$y = 2^{l(x)} \Leftrightarrow y \text{ is a power of } 2 \wedge [y - 1 \leq x \leq 2 \cdot (y-1)],$$
because, if y is a power of 2, $y-1$ is the smallest number with the same length as y, and $2 \cdot (y-1)$ is the largest.
4. Finally,
$$\begin{aligned} x^\frown y = z &\Leftrightarrow x \cdot 2^{l(y)} + y = z \\ &\Leftrightarrow (\exists v < z)(\exists w < z)[v = 2^{l(y)} \wedge x \cdot v = w \wedge w + y = z]. \end{aligned}$$

Finite Sequences

Using the numerical concatenation function $\frown$ we can define numerical properties that reflect properties of words relevant to the operations of elementary formal systems. These include

"***x*** is an initial segment of ***y***,"
"***x*** is a final segment of ***y***,"
"***x*** is a subword of ***y***,"
"***x*** is of the form ***u*** ⇒ ***v***,"

and all quantifiers occurring in the definitions can be bounded for the same reasons as in the four definitions above.

Then for a given EFS it is possible to define arithmetical relations

- $\text{Axiom}(x)$ that reflects "***x*** is an axiom,"
- x subst y that reflects "***x*** gives ***y*** by substitution,"
- x, y modusponens z that reflects "***x***, ***y*** give ***z*** by modus ponens,"

using only bounded quantifiers. But to reflect the property "***x*** is a theorem" we need to state the existence of a *finite sequence*, each term of which is either an axiom or a consequence of earlier terms. Such a sequence is a "proof," and to be a theorem is to be the last term of a proof.

Thus another prerequisite for arithmetizing computation is a device for encoding and decoding finite sequences of positive integers. It is clear that a finite sequence of words $w_1, w_2, \ldots, w_k$ is easily encoded by a single word $*w_1 * w_2 * \cdots * w_k *$ with the help of a new symbol $*$ as a "separator." We can then encode the sequence by a binary numeral and extract information from the corresponding number with the help of the $\frown$ function.

However, there is an arithmetically simpler way to encode a finite sequence of numbers by a single number, due to Gödel (1931). This is by means of the *Gödel β-function*, which is easily defined in terms of the remainder function:

$$\text{rem}(a, b) = r \Leftrightarrow (\exists q < a)(a = bq + r \wedge r < b).$$

The definition of the β-function is by the quantifier-free formula:

$$\begin{aligned}\beta(c, d, i) = x &\Leftrightarrow \text{rem}(c, 1 + (i+1) \cdot d) = x \\ &\Leftrightarrow (\exists q < c)[c = (1 + (i+1) \cdot d) \cdot q + x \wedge x < 1 + (i+1) \cdot d].\end{aligned}$$

Now we can represent any finite sequence of positive integers $x_1, x_2, \ldots, x_n$ as the sequence of values $\beta(c, d, i)$ for suitable c, d, and for

$i = 1, 2, \ldots, n$. This is a consequence of the *Chinese remainder theorem* from elementary number theory, according to which any sequence of positive remainders x_i is attainable when a suitable c is divided by a suitable $1 + (i+1)d$.[4]

EFS-generated Sets are Σ_1^0

Thanks to the β-function, we can express "there is a sequence $x_1, x_2, \ldots, x_n$" by the formula:

$$(\exists c, d, n)(\forall i \le n)[\beta(c, d, i) = x_i].$$

Consequently, "there is a proof" is expressed by the Σ_1^0 formula saying that there is a sequence, each term of which is either an axiom or the consequence of previous terms by substitution or modus ponens:

$$\begin{aligned}(\exists c, d, n)(\forall i \le n)[&\text{Axiom}\beta(c, d, i)\vee\\ &(\exists j < i)(\beta(c, d, j)\text{ subst }\beta(c, d, i))\vee\\ &(\exists j, k < i)(\beta(c, d, j), \beta(c, d, k)\text{ modusponens }\beta(c, d, i))].\end{aligned}$$

And finally, "x is a theorem" is expressed by the Σ_1^0 formula which adds the clause $\beta(c, d, n) = x$, saying "x is the last term in the sequence," to the formula above:

$$\begin{aligned}(\exists c, d, n)[&\beta(c, d, n) = x\wedge\\ &(\forall i \le n)[\text{Axiom}\beta(c, d, i)\vee\\ &\quad(\exists j < i)(\beta(c, d, j)\text{ subst }\beta(c, d, i))\vee\\ &\quad(\exists j, k < i)(\beta(c, d, j), \beta(c, d, k)\text{ modusponens }\beta(c, d, i))]].\end{aligned}$$

Therefore, since an EFS-generated set W of words w corresponds, by definition, to the set of theorems Pw of some EFS, it follows that W is Σ_1^0.

5.9 ARITHMETIZING COMPUTABLE ENUMERATION

In this section we are going to formalize the idea that a nonempty Σ_1^0 set S can be "listed" in a computable fashion. To be precise, we show that

[4] Smullyan takes the concatenation route to finite sequences because he wishes to avoid using number theory. However, it is not clear to me where to draw the "number theory" line. For example, we are using properties of divisors in our definition of "x is a power of 2."

there is a function f whose range is S—so $f(0), f(1), f(2), \ldots$ is a "list" of the members of S—and f is computable in the sense that the relation $f(m) = n$ can be expressed by both a Σ_1^0 and a Π_1^0 formula of PA. This is the arithmetical equivalent of being computably enumerable and having a computably enumerable complement.

Arithmetizing Recursion

The main problem in arithmetizing the definition of f is that the definition is *recursive*; that is, $f(m+1)$ is defined in terms of $f(m)$. So, to allow the arithmetization to go as smoothly as possible, we first explain how to arithmetize (a special case of) recursion.

Definition. A function $F : \mathbb{N} \to \mathbb{N}$ is said to be representable in PA, or *arithmetically representable*, if the relation $F(n) = m$ is equivalent to a formula $\psi(m, n)$ in the language of PA.

Arithmetizing a recursion. *If F is arithmetically representable and f is defined by $f(0) = x_0$ and $f(m+1) = F(f(m))$, then f is arithmetically representable.*

Proof. The idea is that $f(m)$ is the last term in a sequence $x_0, x_1, \ldots, x_m$, where $x_0 = f(0)$ and, for each $i < m$, $x_{i+1} = F(x_i)$. We can express the existence of such a sequence with the help of the Gödel β-function from the previous section. Namely, $x_i = \beta(c, d, i)$ for certain c, d, so the statement that $f(m) = n$ becomes

$$(\exists c, d)[\beta(c, d, 0) = x_0 \wedge (\forall i < m)(\beta(c, d, i+1) \\ = F(\beta(c, d, i))) \wedge \beta(c, d, m) = n].$$

Since β and F are arithmetically representable, we now have an arithmetical formula representing f. □

We note at this stage that the only explicit quantifiers in this formula are the $\exists$ quantifiers in front. So, provided that F is itself Σ_1^0, we will have a Σ_1^0 formula for the relation $f(m) = n$. We will also have a Σ_1^0 formula for the relation $f(m) \neq n$, because this requires only that we change the "$= n$" at the end of the formula to "$\neq n$." Then, since $f(m) = n$ is equivalent to $\neg f(m) \neq n$ we can also express the relation $f(m) = n$ by the negation of the Σ_1^0 formula for $f(m) \neq n$; that is, by a Π_1^0 formula.

Computable Enumeration

We begin with a couple of elementary observations that allow us to work with simple hypotheses.

First, we observe that any Σ_1^0 formula $\exists m_1 \cdots \exists m_k \; \varphi(m_1, \ldots, m_k, n)$, where φ is Σ_0^0, is equivalent to one with a single $\exists$ quantifier, namely

$$\exists m \; \varphi(P_1^k(m), \ldots, P_k^k(m), n),$$

where $P_1^k, \ldots, P_k^k$ are the projection functions for the k-tupling function P^k. All of these functions come from the pairing function P of section 2.4. Thus the k-tupling functions are

$$\begin{aligned} P^2(m_1, m_2) &= P(m_1, m_2), \\ P^3(m_1, m_2, m_3) &= P(m_1, P(m_2, m_3)), \\ P^4(m_1, m_2, m_3, m_4) &= P(m_1, P(m_2, P(m_3, m_4))), \end{aligned}$$

and so on. And when $m = P^3(m_1, m_2, m_3)$, for example, the projection functions are

$$P_1^3(m) = m_1, \quad P_2^3(m) = m_2, \quad P_3^3(m) = m_3.$$

The pairing function P is Σ_0^0 because (by section 2.4)

$$P(x, y) = z \Leftrightarrow 2 \cdot z = 2 \cdot x + (x + y)(x + y + 1),$$

and hence so are its projection functions because

$$P_1(z) = x \Leftrightarrow (\exists y \leq z)[P(x, y) = z], \quad P_2(z) = y \Leftrightarrow (\exists x \leq z)[P(x, y) = z].$$

It follows that all of the functions $P^k, P_1^k, \ldots, P_k^k$ are Σ_0^0, so

$$\exists m \; \varphi(P_1^k(m), \ldots, P_k^k(m), n) \quad \text{is a } \Sigma_1^0 \text{ formula.}$$

Thus, without loss of generality, *we can assume that any Σ_1^0 set S is defined by*

$$n \in S \Leftrightarrow \exists m \; \varphi(m, n),$$

where φ is a Σ_0^0 formula.

Second, if S is nonempty, but finite, we can choose a $\varphi(m, n)$ that holds for infinitely many pairs $\langle m, n \rangle$. Namely, if $S = \{n_1, \ldots, n_l\}$ then

$$n \in S \Leftrightarrow \exists m \varphi(m, n), \quad \text{where} \quad \varphi(m, n) = \exists m[m = m \wedge (n = n_1 \vee \cdots \vee n = n_l)].$$

Computable enumeration of a Σ_1^0 set. *If S is a nonempty Σ_1^0 set, then S is the range of a function g that is both Σ_1^0 and Π_1^0.*

Proof. By the remarks above we can assume that

$$n \in S \Leftrightarrow \exists m\, \varphi(m,n),$$

where φ is a Σ_0^0 formula satisfied by infinitely many pairs $\langle m,n\rangle$. We first consider a function f whose range consists of the numbers $t = P(m,n)$ such that $\varphi(m,n)$. Namely, define f recursively by

$$\begin{aligned} f(0) &= \text{least } t \text{ such that } \varphi(P_1(t),P_2(t)), \\ f(s+1) &= \text{least } t > f(s) \text{ such that } \varphi(P_1(t),P_2(t)). \end{aligned}$$

In other words, f is defined recursively by the equations $f(0) = t_0$ (say) and $f(s+1) = F(f(s))$, where

$$\begin{aligned} F(u) = v &\Leftrightarrow v = \text{least } t > u \text{ such that } \varphi(P_1(t),P_2(t)) \\ &\Leftrightarrow v > u \wedge \varphi(P_1(v),P_2(v)) \wedge (\forall i < v)(i > u \Rightarrow \neg\varphi(P_1(i),P_2(i))). \end{aligned}$$

Thus F has a Σ_0^0 definition and so it follows from arithmetization of recursion that f is both Σ_1^0 and Π_1^0.

Finally, since f has range $\{t : \varphi(P_1(t),P_2(t))\} = \{P(m,n) : \varphi(m,n)\}$ and $P_2(P(m,n)) = n$, it follows that the function

$$g(t) = P_2(f(t))$$

has range $\{n : \exists m\, \varphi(m,n)\}$, as required. □

5.10 ARITHMETIZING COMPUTABLE ANALYSIS

Now that we know the arithmetical meaning of computably enumerable set, computable set, and computable function, we can see how to modify PA axioms to make a system for computable analysis.

First, there should be an axiom (schema) asserting the existence of computable sets. That is, if φ is a property of natural numbers that can be expressed in both Σ_1^0 and Π_1^0 forms, then we have the axiom

$$\exists X\, \forall n\, [n \in X \Leftrightarrow \varphi(n)], \qquad \text{(RCAx)}$$

called the *recursive* (meaning computable) *comprehension axiom*. RCAx is really an axiom schema because there are infinitely many properties

φ expressible in both Σ_1^0 and Π_1^0 forms. Note also that φ can contain set variables other than X. This allows us to apply computable operations to arbitrary sets. For example, RCAx allows us to conclude that if Y is a set, so is the collection Z of even numbers in Y, because

$$n \in Z \Leftrightarrow n \in Y \wedge (\exists m < n)(n = 2 \cdot m)$$

and the condition on the right is both Σ_1^0 and Π_1^0.

Second, we restrict the induction axiom (schema) of PA,

$$[\varphi(0) \wedge \forall n\ (\varphi(n) \Rightarrow \varphi(n+1))] \Rightarrow \forall n\ \varphi(n),$$

to Σ_1^0 formulas φ. We call the latter scheme Σ_1^0 *induction*. The system obtained from PA by restricting induction to Σ_1^0 induction and adding the recursive comprehension axiom is called RCA_0, from the initial letters of "recursive comprehension axiom."[5]

As we have seen in chapter 4, there are many *non*computable sets and functions, so we expect RCA_0 to have limited scope. Indeed, we will see that RCA_0 fails to prove many of the basic theorems of analysis. However, RCA_0 is surprisingly good at proving *equivalences* between important theorems. For example, RCA_0 can prove that the Heine-Borel theorem is equivalent to the extreme value theorem for continuous functions, even though it is unable to prove either of these theorems outright. This makes RCA_0 a good base theory for analysis, because the job of a base theory is to give "elementary" equivalence proofs between theorems that are not themselves "elementary."

Example of a Proof in RCA_0

The proof in the previous section, on the computable enumeration of a Σ_1^0 set, translates into a proof in RCA_0 of the following theorem.

[5]The word "recursive" here is a relic of the time (roughly 1930–1990) when all computable functions were called "recursive." Today the word "computable" is preferred when the general computation concept is meant, and "recursive" is mostly confined to *definitions* (like that in the previous section) where the value of a function is determined by its previous values. But we seem to be stuck with the word "recursive" in the "recursive comprehension axiom."

The subscript 0 in RCA_0 also has a history. Friedman (1975) proposed a system RCA with induction for all arithmetical formulas φ, but Friedman (1976) found that Σ_1^0 induction usually suffices, hence the name change. Σ_1^0 induction is preferred because we want a base system to be as elementary as possible.

Realizing a Σ_1^0 condition by a function. *For any Σ_1^0 condition $\exists m\varphi(m,n)$ there is a function $g:\mathbb{N}\to\mathbb{N}$ such that $\exists m\ [g(m)=n] \Leftrightarrow \exists m\ \varphi(m,n)$.*

Proof. Given $\varphi(m,n)$ we can proceed, as in the previous proof, to write down the definition of a function g whose values are the n such that $\exists m\ \varphi(m,n)$. The function g can be defined by both a Σ_1^0 and a Π_1^0 formula, as we observed, so g exists by the recursive comprehension axiom.

The first step of the previous proof, recursively defining a function f whose values are the numbers $t=P(m,n)$ such that $\varphi(m,n)$, can now be justified by Σ_1^0 induction. Thus we have a proof in RCA_0 that g exists and that $\exists m\ [g(m)=n] \Leftrightarrow \exists m\ \varphi(m,n)$. □

Remember that when RCA_0 proves existence of a function g this means that the set of ordered pairs $\langle n, g(n)\rangle$ is computable. We have *not* proved in RCA_0 that the set $\{n : \exists m\ \varphi(m,n)\}$, the range of g, exists. Indeed, we cannot do this in RCA_0 when $\{n : \exists m\ \varphi(m,n)\}$ is computably enumerable but not computable.

To claim the existence of the set $\{n : \exists m\ \varphi(m,n)\}$ we would need more than recursive comprehension; we would need Σ_1^0 *comprehension.* A system which has Σ_1^0 comprehension, called ACA_0, is the subject of our next chapter.

The Minimal Model of RCA_0

The recursive comprehension axiom means that any model of RCA_0 in which the symbols $0, S(0), SS(0), \ldots$ have their usual interpretation as the natural numbers must contain all the computable sets, since these are precisely the sets definable by both Σ_1^0 and Π_1^0 formulas. So the latter sets are necessarily in any model by the recursive comprehension axiom. It is also sufficient to have only the computable sets in the model, since any set definable from computable sets by a condition that is both Σ_1^0 and Π_1^0 is itself computable.

Thus the *minimal model* of RCA_0 consists of the natural numbers (interpreting the number variables) and the computable sets (interpreting the set variables). It follows that any theorem of RCA_0 must hold in the minimal model. This is why theorems that involve the existence of noncomputable sets cannot be proved in RCA_0.

For example, RCA_0 cannot prove the existence of the range of every function g, because there is a function g that is computable (and hence in the minimal model) with a noncomputable range (hence not in the minimal model).

CHAPTER 6

Arithmetical Comprehension

If we wish to develop analysis in a system based on PA with set variables, the most obvious set existence axiom to use is one called *arithmetical comprehension*. This axiom asserts the existence of a set X of natural numbers for each property φ definable in the language of PA.

More precisely, if $\varphi(n)$ is a property defined in the language of PA plus set variables, but *with no set quantifiers*, then there is a set X whose members are the natural numbers n such that $\varphi(n)$. In symbols,

$$\exists X \forall n [x \in X \Leftrightarrow \varphi(n)]. \qquad (*)$$

Since we assert (*) for all such formulas φ, the arithmetical comprehension axiom is really an axiom schema.

The reason we allow set variables in φ is to enable sets to be defined in terms of "given" sets, as mentioned in section 5.10. The reason we disallow set quantifiers in φ is to avoid definitions in which a set is defined in terms of all sets of natural numbers (and hence in terms of itself).

The system consisting of PA plus arithmetical comprehension (*) is called ACA_0. This system lies at a remarkable "sweet spot" among axiom systems for analysis. It is strong enough to prove all the basic theorems of analysis—as we will see—yet of precisely the same strength as PA in proving theorems of pure number theory (that is, theorems not involving set variables).

Also remarkable is the fact that arithmetical comprehension does not merely imply the basic theorems of analysis. It is actually *equivalent* to some of them, and the equivalences can be proved in the weak system RCA_0 of "computable analysis" introduced at the end of the previous chapter.

6.1 THE AXIOM SYSTEM ACA_0

ACA_0 has the same axioms as PA, except that PA induction is replaced by set variable induction (previously mentioned in section 2.6),

$$\forall X\left[\left[0 \in X \wedge \forall n(n \in X \Rightarrow n+1 \in X)\right] \Rightarrow \forall n(n \in X)\right],$$

and the set existence axiom (schema) is *arithmetical comprehension*:

$$\exists X(n \in X \Leftrightarrow \varphi(n)) \qquad \text{(ACAx)}$$

where $\varphi(n)$ is any formula with no set quantifiers and in which X is not a free variable. In particular, if $\varphi(n)$ has no set variables—so it is a formula of PA—then the set variable induction axiom above holds for the set X of n with property $\varphi(n)$, so

$$[\varphi(0) \wedge \forall n(\varphi(n) \Rightarrow \varphi(n+1))] \Rightarrow \forall n\varphi(n).$$

Thus, in the presence of arithmetic comprehension, set variable induction implies PA induction, and therefore all theorems of PA can be proved in ACA_0. In section 6.8 we explain a more surprising converse fact about ACA_0: its theorems not involving set variables are theorems of PA.

Thus the ability of ACA_0 to prove facts about sets of natural numbers (and hence about real numbers and functions) does not help at all in proving facts about the natural numbers themselves. However, it does enable ACA_0 to prove the basic theorems of analysis, as we will see in the sections below.[1]

The Minimal Model of ACA_0

Since the axioms of ACA_0 include the Peano axioms, any model of ACA_0 includes objects denoted by $0, S(0), SS(0), \ldots$ with the properties of the natural numbers. Conversely, these objects *suffice* to satisfy the Peano axioms among those of ACA_0.

In addition, a model of ACA_0 must include enough *subsets* of the set of objects denoted by $0, S(0), SS(0), \ldots$ to satisfy the arithmetic comprehension schema (ACAx). Since there is an instance of (ACAx) for

[1]The subscript 0 in ACA_0 reflects its history as a weakening of a predecessor system called ACA, in which the formula φ in (ACAx) was allowed to have set quantifiers. ACA does not lie at such a "sweet spot" as ACA_0, because ACA is stronger than PA in proving theorems of pure number theory. We give an example of such a theorem in section 6.6.

each PA formula $\varphi(n)$, these subsets include all the arithmetically definable sets. Conversely, the arithmetically definable sets suffice to satisfy the axioms of ACA_0. They satisfy arithmetic comprehension, even when the defining formula $\varphi(n)$ contains set variables, because a set defined in terms of other arithmetically definable sets is itself arithmetically definable. And they satisfy set variable induction because, as explained above, when the X in the induction axiom denotes an arithmetically definable set, we have an instance of PA induction.

Thus the minimal model of ACA_0 consists of the natural numbers together with all arithmetically definable sets of natural numbers.

6.2 Σ_1^0 AND ARITHMETICAL COMPREHENSION

In this section we prove that arithmetical comprehension follows from a seemingly weaker comprehension axiom: Σ_1^0 *comprehension*:

$$\exists X(n \in X \Leftrightarrow \varphi(n))$$

where $\varphi(n)$ is a Σ_1^0 formula in which X is not a free variable. (Thus any set variables in φ are different from X, and they are not quantified.) What makes this result possible is the fact that the formula $\varphi(n)$ in Σ_1^0 can contain set variables, so sets can be defined *in terms of* previously defined sets.

From Σ_1^0 to arithmetical comprehension. *Each instance of arithmetical comprehension is provable by Σ_1^0 comprehension.*

Proof. We know from section 2.6 that each arithmetical formula is Σ_n^0 for some n. So it suffices to prove, by induction on n, that each Σ_n^0 instance of arithmetical comprehension is provable by Σ_1^0 comprehension.

The base step $n = 1$ is immediate by Σ_1^0 comprehension, so it remains to show how to get from a Σ_k^0 set to a Σ_{k+1}^0 set by Σ_1^0 comprehension.

Let $\varphi(n) = \exists l \forall m \psi(l, m, n)$ be a Σ_{k+1}^0 formula, so $\exists m \neg\psi(l, m, n)$ is Σ_k^0. It follows, by the induction hypothesis, that the set

$$Y = \{\langle l, n\rangle : \exists m \neg\psi(l, m, n)\}$$

is obtainable by Σ_1^0 comprehension. Notice $\langle l, n\rangle \notin Y \Leftrightarrow \forall m \psi(l, m, n)$, so the set

$$X = \{n : \varphi(n)\}$$

is definable by the formula

$$\begin{aligned} n \in X &\Leftrightarrow \exists l \forall m \psi(l, m, n) \\ &\Leftrightarrow \exists l(\langle l, n\rangle \notin Y). \end{aligned}$$

The latter formula is Σ_1^0 with the free set variable Y. Therefore, since Y is obtainable by Σ_1^0 comprehension, so is X. □

Σ_1^0 *Comprehension and the Range of Functions*

The connection between arithmetical comprehension and analysis is established by showing Σ_1^0 comprehension equivalent (in RCA_0) to the following proposition:

Range existence. *The range of any injective function $f : \mathbb{N} \to \mathbb{N}$ exists.*

It is clear that Σ_1^0 comprehension implies range existence, because the range R of f is definable from f by the Σ_1^0 condition

$$n \in R \Leftrightarrow \exists m[f(m) = n].$$

However, the converse is quite subtle. It depends on the arithmetization of recursion we established in section 5.9 and the recursive comprehension axiom of section 5.10 (we cannot expect to get Σ_1^0 comprehension without assuming *some* form of comprehension).

Range existence ⇒ Σ_1^0 comprehension. *If the range of any injective function $f : \mathbb{N} \to \mathbb{N}$ exists, then Σ_1^0 comprehension holds.*

Proof. As we saw in section 5.10, we can prove in RCA_0 the existence of a function whose values satisfy a given Σ_1^0 condition. So, if the range of any function exists, the set of values satisfying any Σ_1^0 condition exists.

That is, Σ_1^0 comprehension holds. □

Σ_1^0 *Induction and Systems Weaker than* ACA_0

In later chapters we will compare ACA_0 with systems having weaker set existence axioms and also the weaker form of induction: Σ_1^0 *induction*:

$$(\varphi(0) \wedge (\varphi(n) \Rightarrow \varphi(n+1))) \Rightarrow \forall n \varphi(n), \quad \text{where } \varphi \text{ is a } \Sigma_1^0 \text{ formula.}$$

In ACA_0, thanks to arithmetical comprehension, Σ_1^0 induction implies induction for any arithmetical φ, but in systems with weaker comprehension this is not the case. The two weaker systems we will be interested in are:

- RCA_0, whose set existence axiom schema asserts only the existence of *computable* sets, and
- WKL_0, whose set existence axiom schema asserts, in addition to the existence of computable sets, the existence of infinite paths in infinite binary trees (the weak Kőnig lemma).

RCA_0 proves only a few basic theorems of analysis, notably the intermediate value theorem. However RCA_0 is strong enough to prove many interesting *equivalences* between set existence axioms and theorems of analysis.

We will prove several such equivalences in the present chapter, and in the next chapter say more about why the proofs can be carried out in RCA_0. In fact, we have already proved one implication in RCA_0: the proof that range existence $\Rightarrow$ Σ^0_1 comprehension. As we observed above and in section 5.10, this is because the proof assumes only the existence of computable sets and Σ^0_1 induction. Roughly speaking, an implication $A \Rightarrow B$ is provable in RCA_0 if the objects asserted to exist in B can be computed from the objects asserted to exist in A.

6.3 COMPLETENESS PROPERTIES IN ACA_0

In this section we see how ACA_0 captures the fundamentals of analysis, by proving arithmetical comprehension equivalent to the following completeness properties of $\mathbb{R}$. These results were announced by Friedman (1976).

1. (Sequential) Bolzano-Weierstrass theorem.
   Any bounded infinite sequence of real numbers has a convergent subsequence.
2. (Sequential) least upper bound principle.
   Any bounded sequence of real numbers has a least upper bound.
3. Cauchy convergence criterion.
   A sequence $x_0, x_1, x_2, \ldots$ is convergent if it has the property that, for any $\varepsilon > 0$, there is an n such that $|x_m - x_n| < \varepsilon$ for all $m > n$.
4. Monotone convergence theorem.
   Any bounded monotonic sequence is convergent.

The proofs reveal the "arithmetical content" of these famous theorems.

To streamline the proofs we assume that real numbers, sequences of real numbers, sequences of closed intervals, and so on, are encoded by sets of natural numbers as explained in chapter 2. We can then assert the existence of various sequences defined by arithmetical conditions (and the objects they may determine—for example, a real number given by a sequence of nested closed intervals whose lengths tend to zero) by arithmetical comprehension. The equivalences will be proved via the implications shown in figure 6.1.

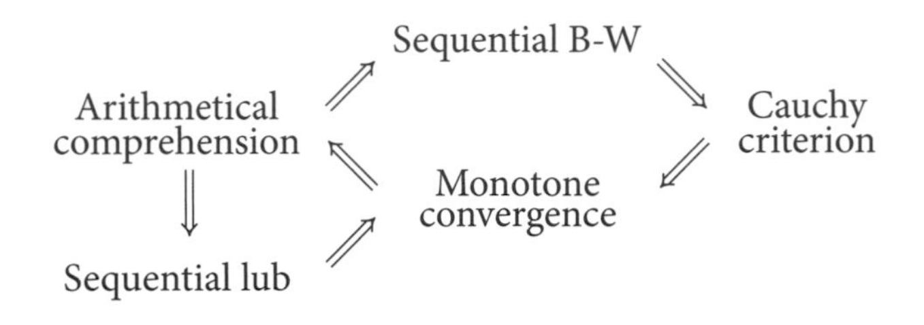

Figure 6.1 : Implications between completeness properties

Arithmetical comprehension ⇒ sequential Bolzano-Weierstrass.

Proof. Suppose that $x_0, x_1, x_2, \ldots$ is a sequence of real numbers, and assume without loss of generality that each $x_i \in [0,1] = I_0$. To find a convergent subsequence we bisect I_0, choose the rightmost half I_1 containing infinitely many x_i, and repeat the process in I_1. This defines a sequence of intervals

$$I_k = [f(k) \cdot 2^{-k}, (f(k)+1) \cdot 2^{-k}],$$

where

$$f(k) = \text{greatest } j < 2^k \text{ such that } j \cdot 2^{-k} \leq x_i \leq (j+1) \cdot 2^{-k}$$
$$\text{for infinitely many } i.$$

This definition is arithmetical, so f (hence the sequence $I_0, I_1, I_2, \ldots$) exists by arithmetical comprehension. Also the I_k are nested and I_k has length 2^{-k}, by a Σ_1^0 induction, so this sequence of intervals defines a real number x. Now from the sequences $x_0, x_1, x_2, \ldots$ and $I_0, I_1, I_2, \ldots$ we define (again, arithmetically)

$$x_{n_k} = \text{first member of } x_0, x_1, x_2, \ldots \text{ in } I_k \text{ with } n_k > n_{k-1}.$$

And this sequence converges, to the point x defined by the intervals I_k. □

Sequential Bolzano-Weierstrass ⇒ Cauchy criterion.

Proof. Suppose that the sequence $x_0, x_1, x_2, \ldots$ satisfies the Cauchy criterion

$$(\forall \varepsilon > 0)\exists n \forall m(m > n \Rightarrow |x_m - x_n| < \varepsilon). \qquad (*)$$

Then the sequence is bounded and hence has a convergent subsequence by Bolzano-Weierstrass.

The limit x of this subsequence $x_{n_0}, x_{n_1}, x_{n_2}, \ldots$ is necessarily the limit of $x_0, x_1, x_2, \ldots$. Because if x_{n_k} is within distance ε of x, then all x_m with $m > n_k$ are within distance 2ε of x by (*), hence the limit of $x_0, x_1, x_2, \ldots$ exists and equals x. □

Cauchy criterion ⇒ monotone convergence.

Proof. This holds because a monotone sequence that does *not* satisfy the Cauchy criterion is unbounded. To see why, suppose we have $\varepsilon > 0$ for which there is no n such that $x_m - x_n < \varepsilon$ for all $m > n$. In that case, for any n there is an $m > n$ with $x_m - x_n > \varepsilon$ (assuming our monotone sequence is increasing). By searching for larger and larger n we find

$$x_n < x_m < x_{n'} < x_{m'} < x_{n''} < x_{m''} < \cdots,$$

with

$$x_m - x_n \geq \varepsilon, \quad x_{m'} - x_{n'} \geq \varepsilon, \quad x_{m''} - x_{n''} \geq \varepsilon, \quad \cdots,$$

so that the sequence $x_0, x_1, x_2, \ldots$ grows beyond all bounds.

Thus a bounded increasing sequence satisfies the Cauchy criterion, and hence is convergent (and similarly for a bounded decreasing sequence). □

Monotone convergence ⇒ arithmetical comprehension.

Proof. It suffices to prove that, for any injective function $f : \mathbb{N} \to \mathbb{N}$, the range of f exists, because this implies Σ_1^0 comprehension (and hence arithmetical comprehension) by section 6.2.

Given an injective function $f : \mathbb{N} \to \mathbb{N}$ we will prove that the range of f exists, in the sense that we can *compute* it from f. We first compute the bounded increasing sequence $c_0, c_1, c_2, \ldots$, where[2]

$$c_n = \sum_{i=0}^{n} 2^{-f(i)}.$$

[2]Readers will notice that this is exactly the same construction that we used in section 4.4 to construct a computable sequence with a noncomputable limit.

Monotone convergence gives the existence of

$$c = \lim_{n\to\infty} = \sum_{i=0}^{\infty} 2^{-f(i)}.$$

And from c we can compute the set of natural numbers n in the range of f because

$$n \in \text{range of } f \Leftrightarrow n\text{th binary digit of } c \text{ is } 1. \qquad \square$$

Arithmetical comprehension ⇒ sequential least upper bound.

Proof. Suppose that $x_0, x_1, x_2, \ldots$ is a bounded sequence, for which we can assume as usual that each $x_i \in [0,1] = I_0$. We let I_1 be the rightmost half of I_0 containing some x_i, and in general

$$I_{k+1} = \text{rightmost half of } I_k \text{ containing some } x_i.$$

Then, as in proving that arithmetical comprehension ⇒ Bolzano-Weierstrass, we find that the sequence $I_0, I_1, I_2, \ldots$ and its common point x exists by arithmetical comprehension.

It follows from the definition of the I_k and x that each $x_i \leq x$ but that, if $y < x$, then $y <$ some x_i. Thus x is the least upper bound of the x_i. □

Sequential least upper bound ⇒ monotone convergence.

Proof. This is because the least upper bound of a monotone increasing sequence is its limit. And a monotone decreasing sequence then has a limit by considering negatives. □

6.4 ARITHMETIZATION OF TREES

In chapter 3 we found that many basic theorems of analysis follow from an infinite bisection process. We noted in section 3.9 that this construction reflects the *weak Kőnig lemma*, stating that an infinite binary tree has an infinite path. The weak Kőnig lemma is so called because it is a special case of the *Kőnig infinity lemma*, stating that an infinite *finitely branching* tree has an infinite path. In the next section we prove the König infinity lemma in ACA_0. The proof is quite simple, but to pave the way for it we have to find suitable encodings of trees by sets of positive integers. There are two natural ways to do this: one specific to binary trees and another for trees in general.

Both ways encode each vertex of the tree by a finite sequence of positive integers, with the "top" vertex being encoded by the empty sequence, and every other vertex v being encoded by a sequence $\langle n_1, \ldots, n_{k-1}, n_k \rangle$, where $\langle n_1, \ldots, n_{k-1} \rangle$ encodes another vertex of the tree (the vertex "above" v). In section 5.8 we saw two ways to encode finite sequences:

- Encoding "letters" $a_1, a_2, a_3, \ldots$ by the dyadic numerals 12, 122, 1222, ... and stringing letters together by the concatenation function $\frown$.
- Encoding a sequence of positive integers $n_1, n_2, \ldots, n_k$ by the values of the Gödel β function $\beta(c, d, i)$ for certain c, d, and $i = 1, 2, \ldots, k$.

Both ways are usable, though the former seems preferable, since concatenation is an integral part of the definition of tree, and we have already defined the function $\frown$ for dyadic numerals.

Indeed, for binary trees we can represent vertices by arbitrary dyadic numerals. The *complete* binary tree C has vertices labelled as shown in figure 6.2 (which is the same as figure 3.7 but with dyadic rather than binary labels). And an arbitrary binary tree B is a subtree of C; that is, B is a set of dyadic numerals with the property that if $u \frown 1 \in B$ or $u \frown 2 \in B$ then $u \in B$.

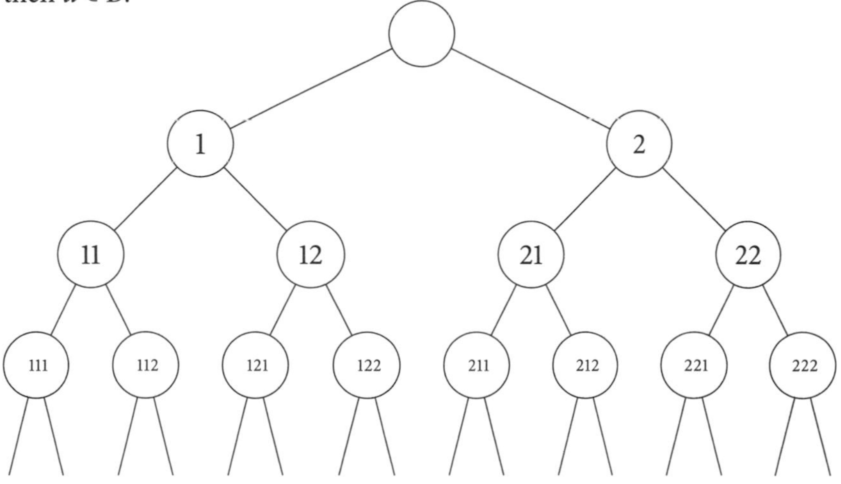

Figure 6.2 : The complete binary tree of dyadic numerals

Definitions. A *tree* is a set T of finite sequences (appropriately coded by dyadic numerals) of positive integers with the property that

$$\langle n_1, \ldots, n_{k-1}, n_k \rangle \in T \Rightarrow \langle n_1, \ldots, n_{k-1} \rangle \in T.$$

T is *finitely branching* if, for each $\langle n_1, \ldots, n_{k-1} \rangle \in T$ there are only finitely many n_k such that $\langle n_1, \ldots, n_{k-1}, n_k \rangle \in T$. Vertex v is an *extension* of vertex u if $u = \langle m_1, \ldots, m_k \rangle$ and $v = \langle m_1, \ldots, m_k, n_1, \ldots, n_l \rangle$ for some $n_1, \ldots, n_l$.

An *infinite path* in a tree T is an infinite sequence $\langle n_1, n_2, n_3, \ldots \rangle$ such that each $\langle n_1, n_2, \ldots, n_k \rangle \in T$.

For example, in the complete binary tree 122 is an extension of 1, and $111 \cdots$ is the leftmost infinite path in the tree.

6.5 THE KŐNIG INFINITY LEMMA

The strength of the Kőnig infinity lemma can be measured by showing that it is equivalent to arithmetical comprehension, as we will do in this section. Their equivalence was announced by Friedman (1976). Surprisingly, the weak Kőnig lemma really is weaker, despite its seemingly similar proof. In both versions one finds an infinite path in a tree by repeatedly choosing one of the finitely many edges that leads into an infinite part of the tree.

Although it is not obvious why, it makes a difference whether the choice is between two edges or an arbitrary finite number. The weak Kőnig lemma will be studied in the next chapter, where we will find its most interesting equivalents.

Kőnig infinity lemma. *If T is an infinite, finitely branching tree, then T has an infinite path.*

Proof. We represent T, as in the previous section, by a set T of finite sequences u of natural numbers. Each member $u \in T$ represents a vertex, so the initial segments of u also belong to T and they represent the vertices above u on the unique path to the top (empty) vertex. The possible successors of $u = \langle n_1, \ldots, n_{k-1} \rangle$ are the sequences $\langle n_1, \ldots, n_{k-1}, n_k \rangle$, where $n_k \in \mathbb{N}$, and we order them by the size of n_k.

The assumption that T is finitely branching means that each $u \in T$ has only finitely many successors. The infinity lemma is proved by recursively choosing successive members of an infinite sequence $\langle n_1, n_2, n_3, \ldots \rangle$ such that each $\langle n_1, \ldots, n_k \rangle \in T$.

Step 1. Let $\langle n_1 \rangle$ be the least of the finitely many vertices below the top (empty) vertex that has infinitely many extensions in T.

Step k. Assuming that $u = \langle n_1, \ldots, n_{k-1} \rangle$ is a vertex with infinitely many extensions in T, choose the successor $\langle n_1, \ldots, n_{k-1}, n_k \rangle$ of u that is least among the finitely many successors of u having infinitely many extensions in T.

Using the arithmetization of trees from the previous section, and arithmetical definability of the property of having infinitely many extensions, we can make a recursive (and hence also arithmetical, by section 5.10) definition of the sequence $\langle n_1, n_2, n_3, \ldots \rangle$. Its existence, and hence that of an infinite path, therefore follows by arithmetical comprehension. □

This shows that *the Kőnig infinity lemma is provable in* ACA_0. Conversely, we have the theorem:

Kőnig infinity lemma ⇒ arithmetical comprehension.

Proof. As in the proof that monotone convergence implies arithmetical comprehension it suffices to prove that the Kőnig infinity lemma implies that the range of any function $f : \mathbb{N} \to \mathbb{N}$ exists, because arithmetical comprehension then follows by sections 5.10 and 6.2.

Thus it remains to show that, given a function $f : \mathbb{N} \to \mathbb{N}$, we can compute the range of f (that is, decide for given n whether $n \in \text{range } f$) with the help of the Kőnig infinity lemma. We do this by computing from f a finitely branching tree T_f whose only infinite path σ is defined by

$$\sigma(i) = \begin{cases} 0 & \text{if } i \notin \text{range } f \\ m+1 & \text{if } f(m) = i. \end{cases}$$

From σ we can decide membership in range f, because

$$i \in \text{range } f \Leftrightarrow \sigma(i) > 0.$$

So if σ exists then range f exists—and we will show that σ exists by the Kőnig infinity lemma.

As explained in the previous section, T_f has a top vertex equal to the empty sequence, and below it vertices that are sequences $\langle m_0, m_1, \ldots, m_k \rangle$ of natural numbers. We decide whether $\langle m_0, m_1, \ldots, m_k \rangle \in T_f$ by the following criterion:

$$\langle m_0, m_1, \ldots, m_k \rangle \in T_f \Leftrightarrow (\forall i, j \le k)[m_i = 0 \Leftrightarrow f(j) \ne i] \text{ and }$$
$$(\forall i \le k)[m_i > 0 \Leftrightarrow f(m_i - 1) = i]. \quad (*)$$

For example, if $\langle 7,5,0,2\rangle \in T_f$ this means that $0,1,3 \in$ range f, because $f(6) = 0, f(4) = 1, f(1) = 3$, and that $2 \notin$ range f unless $f(m) = 2$ for some $m > 3$. More generally, we notice the following:

1. If $\langle m_0, m_1, \ldots, m_k\rangle \in T_f$ then $m_i = 0$ only if $f(m) = i$, if at all, for some $m > k$.
2. If $\langle m_0, m_1, \ldots, m_k\rangle \in T_f$ and $l < k$ then $\langle m_0, m_1, \ldots, m_l\rangle \in T_f$, because if (*) holds for k it also holds for $l < k$. Thus T_f is a tree.
3. Since all quantifiers in (*) are bounded, (*) is decidable from f, so T_f exists by recursive comprehension.
4. T_f is finitely branching, because each vertex $\langle m_0, m_1, \ldots, m_k\rangle \in T_f$ has at most two successors, $\langle m_0, m_1, \ldots, m_k, 0\rangle$ and $\langle m_0, m_1, \ldots, m_k, m+1\rangle$, the latter if $k+1 = f(m)$ for some $m > k$.
5. Each initial segment $\langle \sigma(0), \sigma(1), \ldots, \sigma(k)\rangle$ of σ is in T_f because it satisfies condition (*). Thus T_f is infinite, and hence it contains an infinite path by the Kőnig infinity lemma.

It remains to show that σ is the *only* infinite path in T_f. To do this it suffices to show that, if

$$\langle m_0, m_1, \ldots, m_k\rangle \neq \langle \sigma(0), \sigma(1), \ldots, \sigma(k)\rangle,$$

then all paths through $\langle m_0, m_1, \ldots, m_k\rangle$ eventually terminate. We can assume, without loss of generality, that $m_k \neq \sigma(k)$ and that the preceding terms of both sequences agree. This can happen only if $m_k = 0$ and $f(m) = k$ only for some $m > k$, in which case $\sigma(k) = m+1$. But then all extensions of $\langle m_0, m_1, \ldots, m_k\rangle$ of length greater than m will fail to satisfy condition (*), and hence they will not be in T_f. □

It is worth noting that the tree T_f in the above proof is actually a binary tree, since each vertex has at most two successors. So it might seem that we need appeal only to the *weak* Kőnig lemma about infinite paths in binary trees. However, to *construct* T_f we have to work in the tree of all natural number sequences, because we cannot foresee how large an m may be needed to obtain a given number i as a value $f(m)$. We cannot construct T_f (or anything similar) inside the complete binary tree.

Since arithmetical comprehension implies the Kőnig infinity lemma, and hence the weak Kőnig lemma, all the classical theorems provable from the weak Kőnig lemma are theorems of ACA_0. In the next chapter we see that the best known of these theorems—the sequential Heine-Borel theorem, uniform continuity, and extreme value theorems—can be proved *equivalent* to the weak Kőnig lemma in RCA_0. Hence they are best placed in a weaker system, WKL_0, which has the weak Kőnig lemma as its set existence axiom.

Before leaving ACA_0, however, it is worth mentioning another important area of mathematics which depends on the full strength of arithmetic comprehension.

6.6 RAMSEY THEORY

Ramsey theory is a large area of mathematics concerned with finding "order within disorder" in both finite and infinite structures. It originated in the paper Ramsey (1930), on logic, but has since become part of combinatorics. In this section we will sketch some basic results of Ramsey theory and describe their relation to ACA_0. A finite example which illustrates the kind of "order" discovered by Ramsey theory is the following: *in any group of six people there are either three who all know each other or three who do not know each other.*

To prove this "baby Ramsey theorem" we represent the six people by vertices of a graph, joining two vertices by a black edge if the corresponding people know each other and by a gray edge if they do not. Figure 6.3 shows one such "acquaintance graph."

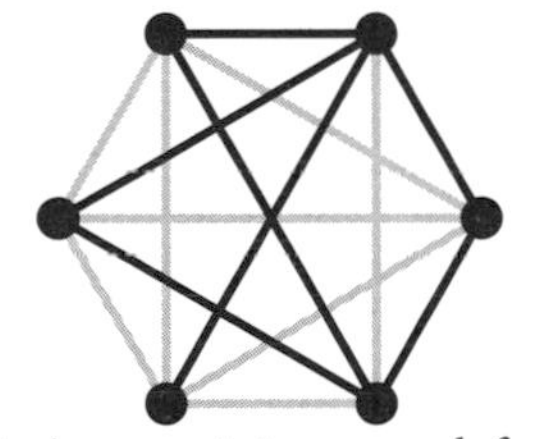

Figure 6.3 : An acquaintance graph for six people

We wish to show that there is always a *monochromatic* triangle (all black or all gray). To see why this triangle exists, consider any vertex v_0 of the graph. It is the endpoint of five different edges, so at least three of these edges must have the same color, say black. Now look at the other ends, v_1, v_2, v_3, of the three edges (figure 6.4).

If any two of v_1, v_2, v_3 are connected by a black edge, this will complete a black triangle with the black edges already present. And if none of the three edges are connected by a black edge, then v_1, v_2, v_3 are the vertices of a gray triangle.

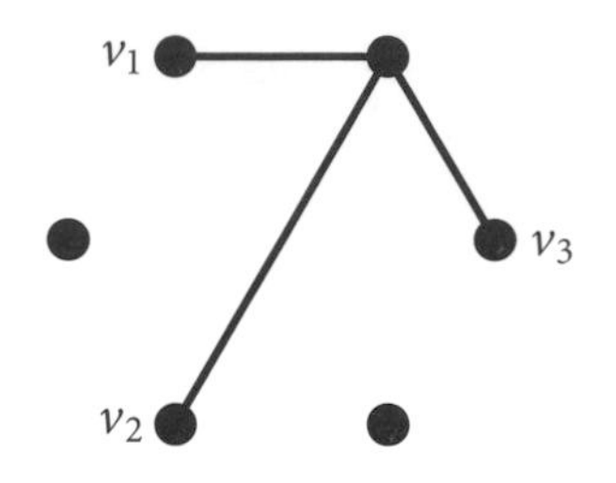

Figure 6.4 : Typical vertex of the graph

Ramsey theory becomes intertwined with reverse mathematics when we consider infinite vertex sets $X \subseteq \mathbb{N}$. The *complete graph* K_X has the vertex set X and an edge between any two members of X. Then an example of what we can prove is:

Infinite Ramsey theorem for pairs. *If the edges of $K_{\mathbb{N}}$ are colored by a finite number of colors, then $K_{\mathbb{N}}$ contains a monochromatic K_X, where X is an infinite subset of $\mathbb{N}$.*

Proof. Choose any $v_0 \in \mathbb{N}$ and consider the infinitely many edges with v_0 at one end. Since there are only finitely many colors, infinitely many edges out of v_0 have the same color. Let X_1 be the set of vertices at the other end of these edges, and choose $v_1 \in X_1$. Again, infinitely many of the edges out of v_1 have the same color. We let X_2 be the set of vertices at the other end of these edges out of v_1, choose $v_2 \in X_2$, and so on.

In this way we obtain an infinite set of vertices $\{v_0, v_1, v_2, \ldots\}$ such that each v_i is connected to $v_{i+1}, v_{i+2}, v_{i+3}, \ldots$ by edges of one color. Also, since there are only finitely many colors, the *same* color occurs for infinitely many of the v_i: call them $x_0, x_1, x_2, \ldots$. Then if $X = \{x_0, x_1, x_2, \ldots\}$ the infinite graph K_X is monochromatic. □

The above theorem is called the Ramsey theorem for pairs because the edges of the graph K_X are essentially pairs of elements of X. We call the theorem RT(2) for short. There is a more trivial Ramsey theorem RT(1), or the "Ramsey theorem for singletons," stating that if the members of $\mathbb{N}$ are colored by finitely many colors then $\mathbb{N}$ has an infinite subset X whose members all receive the same color. RT(1), which also follows from arithmetical comprehension, is known as the *infinite pigeonhole principle*. We actually assumed RT(1) in the proof above, so in effect we proved RT(1) $\Rightarrow$ RT(2).

Likewise, there is a Ramsey theorem RT(3), stating that if all *triples* of members of $\mathbb{N}$ are colored by finitely many colors then $\mathbb{N}$ has an infinite subset X whose triples all have the same color. And RT(2) $\Rightarrow$ RT(3) by a slight elaboration of the argument above. Indeed, by similarly passing from 3 to 4, 4 to 5, and so on, we can prove a Ramsey theorem RT(k) about coloring k-tuples in finitely many colors.

Each Ramsey theorem RT(k) can be proved in ACA_0 and, more interestingly, each RT(k) for $k \geq 3$ is equivalent to arithmetical comprehension over RCA_0. Two remarkable *non*-equivalents of arithmetical comprehension are the following:

1. RT(2), which does not imply arithmetical comprehension, but implies recursive comprehension, and
2. $\forall k\mathrm{RT}(k)$, which implies arithmetical comprehension but is not provable in ACA_0.

For more on the position of Ramsey theorems in reverse mathematics, see Hirschfeldt (2015). $\forall k\mathrm{RT}(k)$ is often called *the* infinite Ramsey theorem, and one of its claims to fame is that it implies the *Paris-Harrington theorem*, a theorem in the language of PA which is not provable in PA.

Paris-Harrington is a modification of the *finite Ramsey theorem* which states that for all $k, l, m \in \mathbb{N}$ there is an $n \in \mathbb{N}$ such that, for any coloring of the k-element subsets of $\{1, 2, \ldots, n\}$ by l colors, there is an m-element subset of $\{1, 2, \ldots, n\}$ whose k-element subsets all have the same color. The finite Ramsey theorem is provable in PA,[3] but if we add the condition that the least element of the m-element subset be greater than m, then the resulting theorem, due to Paris and Harrington (1977), is not provable in PA.

In section 6.8 we show that the "arithmetical" theorems of ACA_0—those not involving set variables—are exactly the same as those of PA. In particular, ACA_0 does not prove the Paris-Harrington theorem, and hence it cannot prove the infinite Ramsey theorem $\forall k\mathrm{RT}(k)$ either. This is an example of a phenomenon called *ω-incompleteness*, where a theory can prove all *instances* $P(0), P(1), P(2), \ldots$ of a certain property $P(k)$, but not $\forall kP(k)$. (The infinite Ramsey theorem is provable in the system ACA mentioned in the footnote in section 6.1, and hence so is the Paris-

[3]Nevertheless, the finite Ramsey theorem remains difficult when it comes to *finding* the numbers that it proves to exist. Even in the case $k = l = 2$ (2-coloring the edges of a graph) the minimum value of n is known only for $m = 2, 3, 4$. For $m = 2$, obviously $n = 2$; for $m = 3$ the value $n = 6$ found in the "baby Ramsey theorem" cannot be bettered; for $m = 4$ it is known that the minimum $n = 18$. For $m = 5$ the minimum n is not yet known.

Harrington theorem. This is how we know that ACA is stronger than ACA_0.)

6.7 SOME RESULTS FROM LOGIC

In his famous lecture *Mathematical Problems*, Hilbert (1902) made a bold statement about the meaning of "existence" in mathematics:

> If contradictory attributes be assigned to a concept, I say that mathematically the concept does not exist. So, for example, a real number whose square is −1 does not exist mathematically. But if it can be proved that the attributes assigned to the concept can never lead to a contradiction by the application of a finite number of logical processes, I say that the mathematical existence of the concept (for example, of a number or a function which satisfies certain conditions) is thereby proved.

The language of predicate logic that we use for arithmetic is an excellent illustration of the principle stated by Hilbert (at greater length): *consistency implies existence.* By analyzing the structure of sentences in arithmetic we will show that: *for any consistent set of sentences in the language of arithmetic there is an interpretation that satisfies them all.*

The nature of this interpretation (which is not necessarily the standard interpretation of arithmetic, but nevertheless quite concrete) will become clearer as we build it. In outline, the process is to find an equivalent of each sentence of a particularly simple form—the so-called *Skolem form*—which has only universal quantifiers and all of them in the front of the sentence, so that satisfying the sentences amounts to satisfying their *instances* for all possible values of their variables. The instances are essentially *Boolean formulas*, that is, formulas whose only logic symbols are Boolean operations. Any inconsistency depends on only finitely many of them, so the problem is reduced to satisfying an infinite set of Boolean formulas when any finite subset can be satisfied.

The latter problem is easily solved by applying the weak Kőnig lemma.

The reduction to Boolean formulas takes place in the following stages.

Stage 1. Reduction of sentences to prenex form.

A sentence σ is in *prenex form* if all of its quantifiers are to the left of the other symbols. We showed how to convert sentences to prenex form in section 2.7.

Stage 2. Making all quantifiers universal.

If not all quantifiers in the prenex form are universal, consider the leftmost $\exists$. Then we have either the form (in which P itself may begin with quantifiers)

$$\exists x P(x), \quad \text{which we replace by } P(a), \text{ for some new constant } a,$$

or else the form

$$\forall x_1 \cdots \forall x_k \exists y P(x_1, \ldots, x_k, y) \quad \text{which we replace by}$$
$$\forall x_1 \cdots \forall x_k P(x_1, \ldots, x_k, f(x_1, \ldots, x_k)), \text{ for some new function symbol } f.$$

The function denoted by f is called a *Skolem function*. If P itself begins with quantifiers we repeat the process in P, and so on, until all $\exists$ quantifiers are removed in favor of constants or function symbols. We then have a universally quantified equivalent of the original formula called its *Skolem form*.

Stage 3. Constructing the domain of Skolem terms.

We have now added new constants $a_1, a_2, a_3, \ldots$ and new function symbols $f_1, f_2, f_3, \ldots$ to the constant 0 and the function symbols S, $+$, and $\cdot$ originally in the language of arithmetic. The terms that can be built from the constants and function symbols are called *Skolem terms*. They include the terms $0, S(0), SS(0), \ldots$ originally present—denoting the natural numbers—plus new Skolem terms such as $S(a_1)$, $S(a_1)+S(a_2)$, and so on, which do not have an obvious interpretation.

We interpret the Skolem terms simply as *themselves*; that is, as strings of symbols. Then the function symbols are interpreted in the natural way as functions on the set of these strings. For example, S is interpreted as the function that sends the string 0 to the string $S(0)$, a_1 to the string $S(a_1)$, and so on.

The interpretation of the relation symbol $=$ remains open at this stage. For example, $a_1 = a_2$ may be true under one interpretation, false under another. We will assign truth values to atomic formulas only at the final stage, when we build a tree of all possible assignments of truth values, and choose those that lie along a certain infinite path in the tree.

Stage 4. Enumerating the atomic instances of the Skolem forms.

Once we have converted a sentence σ to an equivalent Skolem form σ^S, which is of the form

$$\forall x_1 \cdot \forall x_k R(x_1, \ldots, x_k) \quad \text{with } R \text{ quantifier-free,}$$

satisfying σ is equivalent to satisfying all *instances* $R(t_1, \ldots, t_k)$ of σ^S, where $t_1, \ldots, t_k$ are Skolem terms. Since R is quantifier-free, it is a Boolean combination of atomic formulas $t = t'$, where t and t' are Skolem terms. Each of these *atomic instances* of σ^S has two possible interpretations, true or false.

To make a tree of all possible interpretations we enumerate the atomic instances in a list $\alpha_1, \alpha_2, \alpha_3, \ldots$. The list can be made by ordering the atomic formulas by *length*, where a_k and f_k both have length k. With this definition, there are only finitely many formulas of a given length, so we can list them by taking the shortest formulas first, then the next shortest, and so on.

We now have all we need to prove a theorem about satisfiability. It has the corollary that any consistent set of sentences $\sigma_1, \sigma_2, \sigma_3, \ldots$ of arithmetic has an interpretation that satisfies them all. The domain of the interpretation will be the set of Skolem terms constructed above.

Consistency implies satisfiability. *If $\tau_1, \tau_2, \tau_3, \ldots$ are a consistent set of Boolean formulas, then there is an interpretation that satisfies them all.*

Proof. Let $\alpha_1, \alpha_2, \alpha_3, \ldots$ be a list of the atomic parts of the formulas $\tau_1, \tau_2, \tau_3, \ldots$. We first construct the complete binary tree of all interpretations (true or false) of $\alpha_1, \alpha_2, \alpha_3, \ldots$ (figure 6.5).

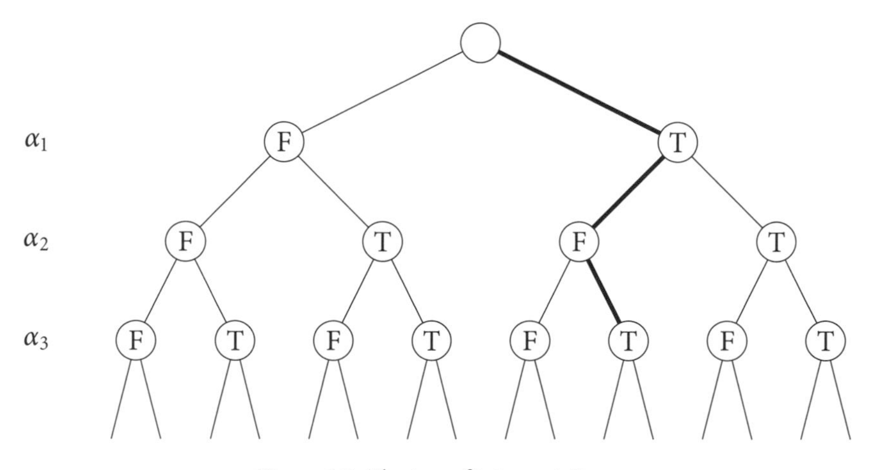

Figure 6.5 : The tree of interpretations

At level 1 (labelled by α_1) there are two vertices, for the two possible interpretations of α_1. Below each of these there are two vertices, for the two interpretations of α_2 for each interpretation of α_1, and so on. Thus the path drawn heavily represents the interpretation $\alpha_1 = \mathrm{T}$, $\alpha_2 = \mathrm{F}$, $\alpha_3 = \mathrm{T}$.

Next we *prune* the complete tree by terminating a path at a vertex on level k if the corresponding values of $\alpha_1, \alpha_2, \ldots, \alpha_k$ falsify one of the formulas τ_i. Since $\tau_1, \tau_2, \tau_3, \ldots$ are consistent, at least one assignment of values to $\alpha_1, \alpha_2, \ldots, \alpha_k$ falsifies no τ_i. Hence the pruned tree has arbitrarily long paths. Then the weak Kőnig lemma gives an infinite path, which necessarily defines an assignment satisfying *all* of $\tau_1, \tau_2, \tau_3, \ldots$. □

Corollary. *If $\sigma_1, \sigma_2, \sigma_3, \ldots$ is a consistent set of sentences in the language of arithmetic, then there is an interpretation that satisfies all of $\sigma_1, \sigma_2, \sigma_3, \ldots$.*

Proof. From $\sigma_1, \sigma_2, \sigma_3, \ldots$ we construct the Skolem forms $\sigma_1^S, \sigma_2^S, \sigma_3^S, \ldots$ and then all their instances, by dropping their quantifiers and replacing their variables by all Skolem terms. The instances make up a consistent set of Boolean formulas, so the theorem above gives an interpretation that makes them simultaneously true.

In making all instances true, the interpretation satisfies the Skolem forms $\sigma_1^S, \sigma_2^S, \sigma_3^S, \ldots$, and hence the original sentences equivalent to them, $\sigma_1, \sigma_2, \sigma_3, \ldots$. □

6.8 PEANO ARITHMETIC IN ACA$_0$

A remarkable feature of ACA$_0$ is that, despite its strength in analysis, it is no stronger than PA in proving theorems about natural numbers. This is quite surprising, in view of the modern history of number theory, where ideas from analysis have become virtually indispensable.[4]

[4]For example, it was thought for about 50 years that analysis was indispensable in the proof of the *prime number theorem*, which states that the number of primes less than n is asymptotic to $n/\ln n$. There was even a premature attempt to develop the "reverse mathematics" of number theory, based on the discovery of Landau that certain theorems can be proved equivalent to the prime number theorem by elementary methods. This led Hardy and Heilbron (1938) to believe that

> it was Landau who first enabled experts to classify the theorems of prime number theory according to their "depths."

However, their claim amounted to the conjecture that the prime number theorem itself did not have an elementary proof—a conjecture which turned out to be false when elementary proofs of the prime number theorem were found by Selberg (1949) and Erdős (1949). To this day we know no "natural" example (that is, arising in number theory, rather than logic) of a theorem of number theory without an elementary proof.

Apparently, analysis lights the path to new discoveries in number theory, but once the path is found it can usually be reconstructed in PA. In the case of ACA_0, we can show that its arithmetic theorems are already in PA as follows.

Suppose that σ_0 is a true sentence in the language of PA not provable in PA. (We know that there are many such sentences, by Gödel's incompleteness theorem.) This means that $\neg\sigma_0$ is consistent with the axioms $\sigma_1, \sigma_2, \sigma_3, \ldots$ of PA, otherwise we could prove σ_0 by deriving a contradiction from $\neg\sigma_0$. It follows, by the theorem in the previous section, that there is an interpretation satisfying all the sentences $\neg\sigma_0, \sigma_1, \sigma_2, \sigma_3, \ldots$. The domain of this interpretation is an extension E of the usual natural numbers $0, S(0), SS(0), \ldots$ by other Skolem terms.

We can extend this model of PA to a model of ACA_0 by adding all the "arithmetically definable" subsets of E, that is, all the sets

$$\{t \in E : \varphi(t)\}, \quad \text{where } \varphi \text{ is a formula of PA with one free variable.}$$

For this to work we need $\varphi(t)$ to have definite truth value for each formula φ and Skolem term t or, equivalently, for the Skolem form φ^S of φ. This is the case, because each φ occurs in an instance σ_i of the induction schema. So its Skolem form φ^S acquires a truth value for each Skolem term in the course of the construction described in the previous section, where truth values are assigned to the atomic parts of all the formulas we wish to satisfy.

Thus there is a model of ACA_0 which satisfies all of $\neg\sigma_0, \sigma_1, \sigma_2, \sigma_3, \ldots$, and therefore σ_0 *is not provable in* ACA_0.

Conversely, if σ is a sentence provable in PA, then σ is provable in ACA_0, since the PA induction schema follows from the ACA_0 induction schema and the other axioms of PA are axioms of ACA_0. In summary: *the purely arithmetic theorems of* ACA_0 *are the same as those of* PA. This theorem is due to Friedman (1976).

It follows in particular that ACA_0 does *not* prove the Paris-Harrington theorem mentioned in section 6.6, since Paris-Harrington is not provable in PA. This tells us why the infinite Ramsey theorem that implies Paris-Harrington (see section 6.6) is not provable in ACA_0, despite its considerable similarity to other Ramsey theorems that ACA_0 can prove.

Relative Consistency of ACA_0

As mentioned in section 4.7, we cannot prove the consistency of PA within PA itself, by Gödel's second incompleteness theorem, and the same applies to any system that contains PA, such as ACA_0. Neverthess, proving the consistency of ACA_0 is not *more* difficult than proving Con(PA), the sentence of PA that expresses the consistency of PA. This is because

$$\begin{aligned} ACA_0 \text{ is consistent} &\Leftrightarrow \text{“}0=1\text{” is not provable in } ACA_0 \\ &\Leftrightarrow \text{“}0=1\text{” is not provable in PA} \\ &\qquad \text{because “}0=1\text{” is a sentence of PA} \\ &\Leftrightarrow \text{Con(PA)}. \end{aligned}$$

We say that ACA_0 is *equiconsistent* with PA. Thus the Hilbert program (described in section 4.7) for proving the consistency of analysis does not fail more badly than the program for proving the consistency of number theory—at least if we interpret "analysis" to be ACA_0 and "number theory" to be PA.

CHAPTER 7

■■■■■

Recursive Comprehension

RCA_0 can be viewed as an axiom system for "computable analysis." Its set existence axiom, called *recursive comprehension*, states the existence of computable sets. (The use of the term "recursive" to mean "computable" is going out of style, but alas it seems set in stone in the name RCA_0 and its set existence axiom.)

The second important feature of RCA_0 is its induction axiom, Σ_1^0 induction, which essentially allows only properties of computably enumerable sets of objects to be proved. This makes RCA_0 a rather weak system—able to prove few theorems outright—yet RCA_0 is surprisingly capable of proving *equivalences* between classical theorems.

This makes RCA_0 a suitable *base theory* for the reverse mathematics of analysis. RCA_0 proves, for example, the equivalence of arithmetical comprehension with the theorems discussed in sections 6.3 and 6.5, and also a new set of equivalences with the weak König lemma. In the present chapter we prove the equivalences between the weak Kőnig lemma and the Heine-Borel, extreme value, and uniform continuity theorems. We also discuss the equivalence of the weak Kőnig lemma with two famous theorems of topology: the Brouwer fixed point and the Jordan curve theorems.

This latter collection of theorems, lying strictly between RCA_0 and ACA_0, establishes the importance of the system WKL_0 whose set existence axiom is the weak Kőnig lemma. Between them, RCA_0, WKL_0, and ACA_0 cover the basic theorems of analysis, and they sort them into three different levels of strength.

7.1 THE AXIOM SYSTEM RCA_0

We briefly introduced RCA_0 in section 5.10. To expand somewhat on what we said there, the axioms of RCA_0 are the non-induction axioms of PA plus, for Σ_1^0 formulas φ,

$$\forall n[\varphi(0) \wedge \forall n(\varphi(n) \Rightarrow \varphi(n+1))] \Rightarrow \forall n\, \varphi(n). \qquad (\Sigma_1^0 \text{ induction})$$

Typically, Σ_1^0 induction is used to prove that all members of a computably enumerated sequence have a certain property. The set existence axiom is, for all Σ_1^0 formulas φ and Π_1^0 formulas ψ,

$$\forall n[\varphi(n) \Leftrightarrow \psi(n)] \Rightarrow \exists X[n \in X \Leftrightarrow \varphi(n)].$$

(Recursive comprehension)

This axiom arises from the result of Post (1944) that a set is computable if and only if it and its complement are computably enumerable, and the fact that computably enumerable sets are Σ_1^0. We reiterate that the formulas φ and ψ can contain set variables, but not set quantifiers. In effect, this enables RCA_0 to prove the existence of any set *computable from* a given set Y; for example, the set Z of even numbers in Y.

An important example of computation from a given set is the diagonal computation by which we proved uncountability of $\mathbb{R}$ in section 1.5.

Uncountability of $\mathbb{R}$. *For any sequence $x_1, x_2, x_3, \ldots$ of real numbers there is a real number $x \neq$ each x_n.*

Proof (in RCA_0). Given the sequence $x_1, x_2, x_3, \ldots$ we compute x by the rule

$$n\text{th decimal digit of } x = \begin{cases} 1 & \text{if } n\text{th decimal digit of } x_n \neq 1 \\ 2 & \text{if } n\text{th decimal digit of } x_n = 1. \end{cases}$$

Thus x exists by recursive comprehension, each digit of x is either 1 or 2, and x differs from x_n in the nth decimal place. Since a decimal expansion whose digits are each 1 or 2 is unique, this makes $x \neq$ each x_n. □

However, as mentioned in section 5.10, RCA_0 has a minimal model in which all the sets are computable. So when a computable set Y is related to a *non*computable set Z, RCA_0 will not be able to prove the existence of Z, since the minimal model will contain Y but not Z. For example, each computable function is in the minimal model (as a set of ordered pairs), but not the range of each computable function, since the range can be noncomputable. As we saw in section 5.10, this is why RCA_0 cannot prove "the range of every function exists." Similar arguments enable us to show that some important theorems of analysis are not provable in RCA_0 because they imply the existence of noncomputable sets. In particular, RCA_0 is too weak to prove the least upper bound principle, the Heine-Borel theorem, or the Bolzano-Weierstrass theorem.

The weakness of RCA_0 is a plus, however, when we want a base theory to prove *equivalences* between stronger axioms and various theorems—because RCA_0 is strong enough to prove many equivalences. For example, the proofs in the previous chapter that the arithmetic comprehension axiom is equivalent to (various statements of) completeness of $\mathbb{R}$, the Kőnig infinity lemma, and the Bolzano-Weierstrass theorem can be carried out in RCA_0. In this chapter we use RCA_0 as base theory in which to prove equivalents of the weak Kőnig lemma.

7.2 REAL NUMBERS AND CONTINUOUS FUNCTIONS

In the above proof that $\mathbb{R}$ is uncountable we assumed that real numbers are given by decimal expansions, and we used the computability of the decimal expansion of the diagonal number x to prove its existence in RCA_0. This is indeed the classical concept of computable number, introduced by Turing (1936). But when doing analysis in RCA_0 it is convenient to use a slightly different real number concept, better suited to the processes of analysis.

Definition. A *real number* is a nested sequence of closed rational intervals

$$[a_1, b_1] \supseteq [a_2, b_2] \supseteq [a_3, b_3] \supseteq \cdots$$

such that $b_n - a_n \to 0$.

The nested interval concept of real number gets around the difficulties RCA_0 has with convergent sequences. The definition does not say the nested sequence has to be computable, but of course to prove in RCA_0 that a real number *exists* we have to exhibit a computation, and then invoke recursive comprehension. To do this it suffices to give a *computable enumeration* of the intervals $[a_n, b_n]$, because the *sequence* is the set of pairs $\langle n, [a_n, b_n]\rangle$ and this set is computable. (Namely, for each n run the enumeration until the nth term appears.)

We note that any computable sequence of intervals $[a_n, b_n]$ with a single common point determines a real number x in the above sense, because we can compute from it the *nested* sequence

$$[a'_1, b'_1] \supseteq [a'_2, b'_2] \supseteq [a'_3, b'_3] \supseteq \cdots$$

where $[a'_n, b'_n] = [a_1, b_1] \cap \cdots \cap [a_n, b_n]$. The sequence $[a'_n, b'_n]$ has the same common point as the sequence $[a_n, b_n]$, so necessarily $b'_n - a'_n \to 0$.

One difficulty we cannot avoid is that, in general, *it is impossible to know whether a real number x equals* 0. When x is defined by the sequence of intervals $[a_n, b_n]$ then $x = 0$ if and only if $a_n \leq 0 \leq b_n$ for all n, and we cannot observe the whole infinite sequence. But if $x > 0$ we will observe this fact at some finite stage, because it will eventually happen that $a_n > 0$. Similarly, when $x < 0$ we will eventually observe this fact. This limited knowledge is enough for some important purposes, such as proving the intermediate value theorem, as we will see in the next section.

Defining real numbers by intervals also helps the study of continuous functions in RCA_0, where we encode such functions by rational intervals as in section 2.5. We want to say that a continuous function f is given by pairs of rational intervals $\langle(c, d), (a, b)\rangle$ such that $f((c, d)) \subseteq (a, b)$, and that there are "enough" intervals to determine $f(x)$ for each real number x in the domain of f. The problem now is that f itself is not given, so we cannot define the set of pairs in terms of it. Instead, we seek some simple and natural conditions on a set of pairs, which ensure that they encode a continuous function f on some domain.

Definitions. A set of ordered pairs $\langle(c, d), (a, b)\rangle$ of rational intervals is called the *code of a continuous function* f, code(f),

1. if $\langle(c, d), (a, b)\rangle \in \mathrm{code}(f)$ and $\langle(c, d), (a', b')\rangle \in \mathrm{code}(f)$ then (a, b) and (a', b') intersect;
2. if $\langle(c, d), (a, b)\rangle \in \mathrm{code}(f)$ and $(c', d') \subseteq (c, d)$ then we have $\langle(c', d'), (a, b)\rangle \in \mathrm{code}(f)$; and
3. if $\langle(c, d), (a, b)\rangle \in \mathrm{code}(f)$ and $(a, b) \subseteq (a', b')$ then we have $\langle(c, d), (a', b')\rangle \in \mathrm{code}(f)$.

A real number x is in the *domain* of f if the intervals (c, d) include a nested sequence (c_n, d_n) with single common point x such that each

(c_n, d_n) is paired with an (a_n, b_n), and the sequence of the (a_n, b_n) has a single common point, which we call $f(x)$.

It follows from these definitions that if a whole interval $(c, d) \subseteq \text{domain}(f)$ then in fact $f((c, d)) \subseteq (a, b)$ for each pair $\langle (c, d), (a, b) \rangle \in \text{code}(f)$. We prove existence of a continuous function f by exhibiting a computation of the set code(f) and appealing to recursive comprehension.

Given code(f) and a point x in domain(f) we can *compute* $f(x)$ from them as follows. We are given x as a sequence (c_n, d_n) so, by listing the (c_n, d_n) alongside all (c, d) occurring in code(f), we can find all intervals (a, b) paired with each (c_n, d_n). These (a, b) become arbitrarily small, because $f(x)$ exists, so for each k we can find a pair $\langle (c_{n_k}, d_{n_k}), (a_{n_k}, b_{n_k}) \rangle$ with $b_{n_k} - a_{n_k} < 1/k$. Since (c_{n_k}, d_{n_k}) is a subsequence of (c_n, d_n) its single common point is x. The computable sequence (a_{n_k}, b_{n_k}) includes the point $f(x)$ by property 1, and *only* this point because $b_{n_k} - a_{n_k} < 1/k$. Thus $f(x)$ is computable from x.

7.3 THE INTERMEDIATE VALUE THEOREM

The unavoidable difficulty with real numbers means that in general we cannot know that $f(x) = 0$, for given x and continuous function f. But when $f(x) > 0$ or $f(x) < 0$ we can eventually observe this fact, which enables us to save the proof of the intermediate value theorem. A naive attempt to prove it, following the classical argument in section 3.3, goes like this.

Given $f(0) < 0$ and $f(1) > 0$, calculate $f(1/2)$. If $f(1/2) = 0$ we are done. If not, then f passes from negative to positive values in either $[0, 1/2]$ or $[1/2, 1]$. Let I_1 be the half of [0,1] in which this happens, and repeat the argument in I_1. Either $f = 0$ at the midpoint of I_1, or there is a half I_2 of I_1 in which f passes from negative to positive values. And so on. This process either finds a midpoint of some interval at which $f = 0$, or else it generates a nested sequence $I_1 \supset I_2 \supset \cdots$ of intervals, on each of which f passes from negative to positive values. But then $I_1, I_2, \ldots$ have a single common point x, at which we necessarily have $f(x) = 0$.

There are two problems with this attempt.

1. If $f(c) = 0$ at some midpoint c we will not necessarily observe this fact, no matter how long we spend computing $f(c)$.

2. If $f(c) < 0$ or $f(c) > 0$ at some midpoint c we will *eventually* observe this fact, but we do not know how long we have to wait.

Nothing can be done about problem 1, so we have to do our best with problem 2. This is the kind of difficulty that computability theorists love! To deal with it we start more and more computations, and wait for the results. In this case, we wait for points c where $f(c) < 0$ or $f(c) > 0$ to show up, and use them as long as they are "sufficiently close" to the middle of the interval.

Intermediate value theorem. *If f is continuous on $[0,1]$ and $f(0) < 0$ and $f(1) > 0$ then $f(c) = 0$ for some c in $[0,1]$.*

Proof. If $f = 0$ on a whole subinterval of [0,1] then $f(c) = 0$ for some rational (and hence computable) point c, which exists by recursive comprehension.

Otherwise, compute values of f in stages as follows.

Stage 1. Do one step in the computation of $f(1/2)$.
Stage s. Do s steps in the computation of $f(r/2^s)$, for $r = 1, 2, \ldots, 2^s - 1$. Thus stage s continues all the computations in progress at stage $s-1$, as well as starting new computations at points half way between the points where computations have already been started.

In this way we will eventually find each point of the form $x = r/2^s$ at which $f(x) > 0$ or $f(x) < 0$. And *every subinterval of* $[0,1]$ *contains such points*, since we now assume that there is no subinterval on which f is constantly zero. This fact allows us to fix the naive attempt above, by looking at "midintervals" instead of midpoints.

We begin with an interval in the middle of [0,1], say the middle third, and let c_1 be the first point x found in it (during the above computation) for which $f(x) > 0$ or $f(x) < 0$. Then f passes from negative to positive values on either $[0, c_1]$ or $[c_1, 1]$. Let I_1 be the subinterval on which this happens, and look at the middle third of I_1. We similarly find a subinterval I_2 of I_1, between a point c_2 in the middle third of I_2 and one of its endpoints, on which f passes from negative to positive values.

Continuing in this way we obtain (by Σ_1^0 induction) a nested sequence of closed intervals $I_1 \supset I_2 \supset I_3 \supset \cdots$ such that f passes from negative to positive values on each I_n. Also, each I_n is at most 2/3 the length of its predecessor, so the length of I_n tends to zero, and hence the sequence $I_1, I_2, I_3, \ldots$ defines a real number, c, by recursive comprehension.

It is clear from the continuity of f that $f(c) = 0$. □

An important special case of the intermediate value theorem is where f is an odd-degree polynomial with real coefficients, say

$$f(x) = x^n + a_{n-1}x^{n-1} + \cdots + a_1x + a_0.$$

In that case $f(x) < 0$ for large negative x and $f(x) > 0$ for large positive x, so there is an interval I_0 on which f passes from negative to positive values. Then the proof above goes through with I_0 in place of [0,1]; indeed we can rule out the possibility that f is constantly zero on a subinterval. We can then use the argument of Gauss (1816), which reduces an arbitrary polynomial equation to one of odd degree, to prove in RCA_0:

Fundamental theorem of algebra. *For any polynomial f with real coefficients, the equation $f(x) = 0$ has a solution in the complex numbers.* □

7.4 THE CANTOR SET REVISITED

To study the weak Kőnig lemma we revisit binary trees and their connection with the Cantor set, C, introduced in section 3.8. There we saw that C consists of the points that remain in [0,1] after removal of the following open intervals, which we will call the *C-complementary* intervals:

$$\left(\tfrac{1}{3}, \tfrac{2}{3}\right)$$

$$\left(\tfrac{1}{9}, \tfrac{2}{9}\right) \qquad \left(\tfrac{7}{9}, \tfrac{8}{9}\right)$$

$$\left(\tfrac{1}{27}, \tfrac{2}{27}\right) \quad \left(\tfrac{7}{27}, \tfrac{8}{27}\right) \quad \left(\tfrac{19}{27}, \tfrac{20}{27}\right) \quad \left(\tfrac{25}{27}, \tfrac{26}{27}\right)$$

....

We also saw that the points of C correspond to infinite paths in a complete binary tree whose vertices at level n are the *closed* intervals that remain after removing the intervals in row n of the list of C-complementary intervals above. Figure 7.1 shows these intervals again, in their successive levels.

We now "expand" each black closed interval to a gray *open* interval by increasing its length by one third at each end (but omitting the endpoints). For example, the intervals $\left[0, \tfrac{1}{3}\right], \left[\tfrac{2}{3}, 1\right]$ at level 1 of figure 7.1 expand to $\left(-\tfrac{1}{9}, \tfrac{4}{9}\right), \left(\tfrac{5}{9}, \tfrac{10}{9}\right)$. The expanded intervals at each level cover C, so we call them *C-covering* intervals. The smallness of the expansion ensures that any two C-covering intervals not on the same path are dis-

Figure 7.1 : [0,1] after removing the C-complementary intervals

joint. Figure 7.2 shows the first four levels of C-covering intervals, superimposed on the vertices of the tree whose infinite paths represent the members of C.

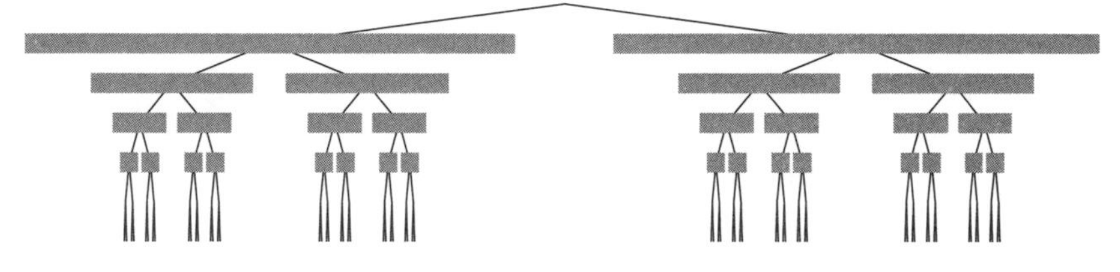

Figure 7.2 : C-covering intervals

Together, the C-complementary and C-covering intervals cover [0,1]. So, according to the Heine-Borel theorem, a finite subset of them also covers [0,1]. This enables us, by choosing C-covering intervals to suit a given binary tree, to use the Heine-Borel theorem to prove finiteness results about trees. In particular, we can show that the Heine-Borel theorem implies the weak Kőnig lemma, and that the implication is provable in RCA_0.

7.5 FROM HEINE-BOREL TO WEAK KŐNIG LEMMA

The classical proof of the Heine-Borel theorem in section 3.5 suggests that it follows from the weak Kőnig lemma, and we might also hope to prove the converse. However, we have to proceed a little differently to prove an equivalence between these two theorems in RCA_0. We have to use the sequential form of Heine-Borel, and we have to make the proofs as computable as possible. These results, and the related ones on uniform continuity, are due to Simpson and we have mostly followed the proofs in Simpson (2009).

Our first aim is to prove that sequential Heine-Borel implies the weak

König lemma in the following form (logically equivalent to the usual form): *if T is a binary tree with no infinite path, then T is finite*. Our approach is to view T as a subtree of the complete binary tree B whose infinite paths are the members of C, and to associate certain C-covering intervals with T. Then we can use the Heine-Borel theorem to deduce that T is finite.

We pass from T to a set of C-covering intervals using vertices called *fallen leaves* of T, defined below, on which we place the corresponding C-covering intervals. Figure 7.3 shows an example, with the edges of T drawn extra thick and C-covering intervals placed on the fallen leaves of T.

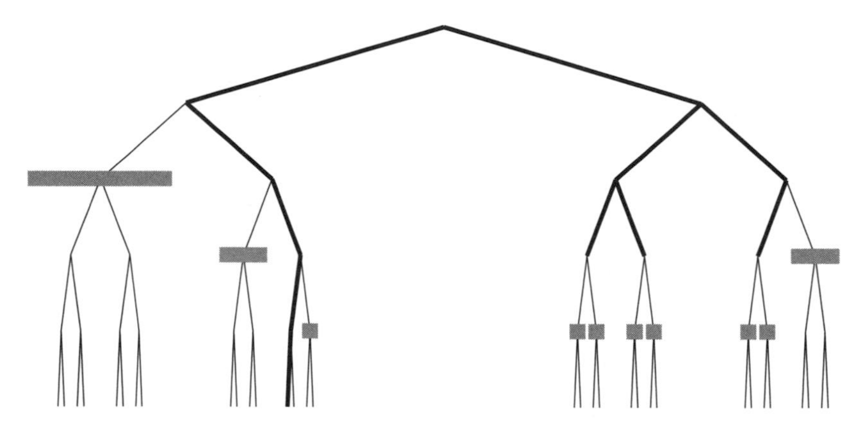

Figure 7.3 : Covering the fallen leaves of a tree

Definition. A vertex v of B is called a *fallen leaf* of T if $v \notin T$ but $u \in T$, where u is the vertex above v.

Two remarks are immediate from this definition:

- If T has no infinite path, then the C-covering intervals that cover fallen leaves of T cover all of C—because an uncovered point corresponds to an infinite path, which T does not have.
- The C-covering intervals corresponding to distinct fallen leaves of a tree T are disjoint—because C-covering intervals overlap only if they lie on the same path, and one fallen leaf cannot lie on the path to another, since a fallen leaf terminates all paths below it.

A more substantive result about the paths not covered is the following:

Computable infinite path lemma. *If T is an infinite computable tree with a finite number of fallen leaves, then T has an infinite computable path.*

Proof. Since T has only finitely many fallen leaves, there is a level n below which no fallen leaves occur. Let v_1 be the leftmost vertex of T below level n, which must exist since T has infinitely many vertices.

Since there is no fallen leaf below v_1, at least one of the two vertices of B immediately below v_1 is in T. Let v_2 be the leftmost of the two, and repeat the argument at v_2. Again, at least one of the two vertices of B below v_2 is in T. Let v_3 be the leftmost of them, and so on.

In this way we can *compute* (from knowledge of the members of T) an infinite path $v_1, v_2, v_3, \ldots$ in T. □

Heine-Borel implies the weak Kőnig lemma. *If T is a tree with no infinite path, then T is finite.*

Proof. If T is a tree with no infinite path, consider the set of C-covering intervals placed on fallen leaves of T. Then, by the first remark above, this set of intervals covers C. Therefore, by combining these intervals with the C-complementary intervals we get a covering of [0,1] by open intervals. The Heine-Borel theorem says that finitely many of these intervals also cover [0,1]. In particular, we get finitely many C-covering intervals, placed on fallen leaves of T, that cover C.

Next, since the intervals placed on distinct fallen leaves are disjoint, by the second remark above, it follows that T has only finitely many fallen leaves. But then, if T is infinite, it follows from the computable infinite path lemma and recursive comprehension that T has an infinite path, contrary to hypothesis. Thus T is finite. □

The implication just proved can be established in RCA_0, thanks to the computability of the path from the tree in the lemma. The implication actually requires only a special case of Heine-Borel, where the covering intervals form a *sequence*. (The set of C-complementary and C-covering intervals can be written in a sequence by listing them level by level.) Also, we can assume that the endpoints of the intervals are *rational*. This is the natural version of Heine-Borel to use, but to make it clear that it is not the general (and classical) version we call it the *sequential Heine-Borel theorem*. Like the classical Heine-Borel theorem, the sequential Heine-Borel theorem is not provable in RCA_0. The counterexample in section 4.5 already shows this.

Thus the proof above shows that the sequential Heine-Borel theorem implies the weak Kőnig lemma in RCA_0. In the next section we will show that the converse implication holds in RCA_0. As mentioned when we first proved the Heine-Borel theorem in section 3.5, some modification of the argument is needed in order to pass "computably" from the weak Kőnig lemma to Heine-Borel.

7.6 FROM WEAK KŐNIG LEMMA TO HEINE-BOREL

The sequential Heine-Borel theorem says: if a sequence $(a_1, b_1), (a_2, b_2)$ $\ldots$ of rational intervals covers [0,1] then some initial segment of the sequence $(a_1, b_1), (a_2, b_2), \ldots, (a_n, b_n)$ also covers [0,1]. We will prove this theorem by deducing it from the weak Kőnig lemma. The proof is somewhat similar to the construction of the Cantor set C, building a binary tree whose vertices at level n are closed intervals that remain (at least in part) when the intervals $(a_1, b_1), (a_2, b_2), \ldots, (a_n, b_n)$ are removed from [0,1].

Weak Kőnig lemma implies sequential Heine-Borel. *If (a_i, b_i) are rational intervals such that the infinite sequence $(a_1, b_1), (a_2, b_2), \ldots$ covers $[0, 1]$ then, for some n, $(a_1, b_1), (a_2, b_2), \ldots, (a_n, b_n)$ also covers $[0, 1]$.*

Proof. Given the sequence $(a_1, b_1), (a_2, b_2), \ldots$ we build a tree T whose vertices at level n are closed subintervals of [0,1]; namely, the subintervals of the form $\left[\frac{m}{2^n}, \frac{m+1}{2^n}\right]$ not completely covered by

$$(a_1, b_1), \quad (a_2, b_2), \quad \ldots, \quad (a_n, b_n).$$

For example, when the first three intervals (a_i, b_i) are $\left(\frac{1}{3}, \frac{4}{3}\right)$, $\left(-\frac{1}{8}, \frac{1}{16}\right)$, and $\left(\frac{5}{32}, \frac{7}{32}\right)$ the first three levels of T are as shown in figure 7.4. (The covering intervals are drawn in white, so they erase the parts of [0,1] they cover.)

The corresponding vertices of T (below the top vertex [0,1]) are

$$\left[0, \tfrac{1}{2}\right] \text{ on level 1 because } \left[\tfrac{1}{2}, 1\right] \text{ is covered by } \left(\tfrac{1}{3}, \tfrac{4}{3}\right),$$

$$\left[0, \tfrac{1}{4}\right], \left[\tfrac{1}{4}, \tfrac{1}{2}\right] \text{ on level 2 because neither is covered by } \left(\tfrac{1}{3}, \tfrac{4}{3}\right), \left(-\tfrac{1}{8}, \tfrac{1}{16}\right),$$

$$\left[0, \tfrac{1}{8}\right], \left[\tfrac{1}{8}, \tfrac{1}{4}\right], \left[\tfrac{1}{4}, \tfrac{3}{8}\right] \text{ on level 3 because } \left[\tfrac{1}{8}, \tfrac{1}{4}\right] \text{ is not covered by } \left(\tfrac{5}{32}, \tfrac{7}{32}\right).$$

There is an edge of T from each subinterval at level n to each half of it that remains at level $n + 1$. Thus T is a binary tree.

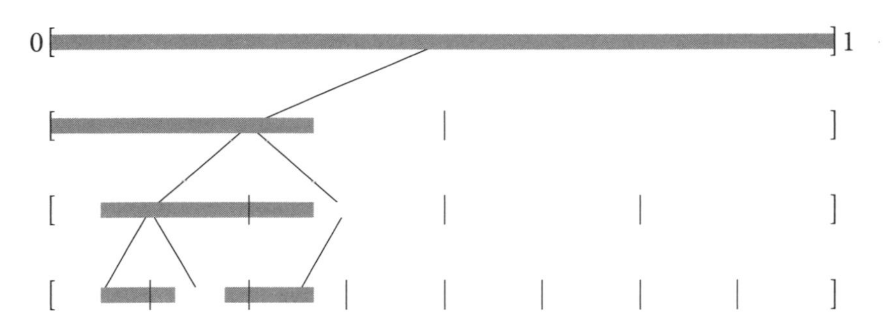

Figure 7.4 : The tree of incompletely covered subintervals

Since $(a_1, b_1), (a_2, b_2), \ldots$ cover [0,1], each $x \in [0,1]$ falls into some (a_i, b_i). Indeed, for n sufficiently large, the one of the 2^n subintervals that contains x falls inside some (a_i, b_i), and so do the subintervals to its left and right. At the stage when this happens none of these three[1] subintervals are put in T, thereby terminating all paths in T that lead to x. Therefore, T has no infinite path, and hence T is finite by the weak Kőnig lemma.

If n is the first level that includes no vertices of T, then it follows from the definition of T that $(a_1, b_1), (a_2, b_2), \ldots, (a_n, b_n)$ covers $[0,1]$. □

In the above proof the tree T is obviously computable from the sequence $(a_1, b_1), (a_2, b_2), \ldots$, so the proof can be carried out in RCA_0, like the proof of the converse result in the previous section. Thus, although neither the weak Kőnig lemma nor the sequential Heine-Borel theorem is provable in RCA_0 (as we saw in section 4.5) their *equivalence* is. In this sense, the weak Kőnig lemma is the "right axiom" to add to RCA_0 in order to prove the sequential Heine-Borel theorem.

7.7 UNIFORM CONTINUITY

In section 3.7 we defined a function f to be *uniformly continuous* on a set S if, for all x, y in S, and all $\varepsilon > 0$, there is a $\delta > 0$ such that

$$|x - y| < \delta \Rightarrow |f(x) - f(y)| < \varepsilon.$$

As we mentioned at the time, uniformity means that δ depends only on ε, not on x or y. It is now convenient to let δ be explicitly a *function* of

[1]This takes care of a worry when x is the endpoint of an interval, in which case two different infinite paths in T could lead to x.

ε, and also to let the "arbitrarily small" $\varepsilon > 0$ be explicitly 2^{-n}, where n is a positive integer. We can then express the dependence of δ on ε by a function h of positive integers, called a *modulus of uniform continuity*, by setting $\delta = 2^{-h(n)}$ when $\varepsilon = 2^{-n}$. Then we make the following:

Definition. A function f on a set S is *uniformly continuous with modulus* h if, for all $x, y \in S$ and all $n \in \mathbb{N}$,

$$|x - y| < 2^{-h(n)} \Rightarrow |f(x) - f(y)| < 2^{-n}.$$

The classical theorem that any continuous function on [0,1] is uniformly continuous now has the following counterpart in RCA_0.

Weak Kőnig lemma implies uniform continuity. *The weak Kőnig lemma implies that any continuous function on* $[0,1]$ *has a modulus of uniform continuity.*

Proof. We skip the details. But the idea is to combine the proof in RCA_0 that the weak Kőnig lemma implies sequential Heine-Borel, from the previous section, with the proof that Heine-Borel implies uniform continuity in section 3.7. The latter proof can also be carried out "computably," leading to a proof in RCA_0. □

More importantly, we can show in RCA_0 that uniform continuity implies the weak Kőnig lemma, so the weak Kőnig lemma is the "right axiom" to prove it.

Uniform continuity implies weak Kőnig lemma. *Uniform continuity of continuous functions on* $[0,1]$ *implies the weak Kőnig lemma.*

Proof. We will prove that if the weak Kőnig lemma fails, then there is a continuous function on [0,1] that is not uniformly continuous. Suppose then that T is an infinite binary tree with no infinite path. From T we will compute a continuous function f on [0,1] that is not uniformly continuous—in fact, f is *unbounded*, because the values of f include the lengths of all paths in T.

By finding the vertices of T level by level, we can enumerate its *fallen leaves* level by level, together with the corresponding C-covering intervals described in sections 7.4 and 7.5. At the same time, we can enumerate the C-complementary intervals of section 7.4 down to the same level. At stage n we define f for x in the union U_n of the intervals (of both kinds) listed down to level n, extending the definition of f at stage $n-1$ on each new interval that appears so as to make f piecewise linear, hence continuous, on U_n.

The definition of f is largely arbitrary; the only important requirement is the following: *when a new C-covering interval I appears at stage n, $f(x)$ is defined to be n for at least one point x in I.* This is always possible, because the C-complementary intervals cover no points of C, so at stage n they leave uncovered a whole *interval* containing points of C. Consequently, when I first appears, f is undefined on a whole subinterval J of I, which is contained in [0,1] since the C-complementary intervals all lie in [0,1]. This allows us to continuously extend f to J so as to obtain $f(x) = n$ for at least one $x \in J$.

It follows that f is unbounded on [0,1], taking the value n for arbitrarily large values of n. This is because T is infinite, but all its paths are finite, so fallen leaves occur at arbitrarily deep levels n. And f is continuous, being continuous at each point $x \in [0, 1]$: indeed, each $x \in [0, 1]$ falls into the open set U_n at some stage, at which stage f becomes defined and continuous at x. □

Corollary. *The extreme value theorem implies the weak Kőnig lemma.*

Proof. The above proof also shows that, if the weak Kőnig lemma fails, then so does the extreme value theorem, because f is unbounded and continuous on [0,1]. Thus the extreme value theorem implies the weak Kőnig lemma. □

The proof of the corollary can be carried out in RCA_0, and so can that of its converse—weak Kőnig lemma implies the extreme value theorem—by attending to computability in the classical proof (section 3.6). We do the converse implication in the next section. Thus the weak Kőnig lemma is also the "right axiom" to prove the extreme value theorem. Not surprisingly, considering the close relation between uniform continuity and Riemann integrability, the weak Kőnig lemma is also the "right axiom" to prove Riemann integrability of continuous functions. For more details on these, and other theorems provably equivalent to the weak Kőnig lemma in RCA_0, see Simpson (2009), pp. 134–135.

7.8 FROM WEAK KŐNIG TO EXTREME VALUE

We begin, as in the classical proof of the extreme value theorem (section 3.6), by supposing that we have an unbounded continuous function on [0,1]. We seek a contradiction by narrowing the region on which f is unbounded, by repeated bisection. But now we have to pursue the place where f is unbounded by a computable process, so we compute a *tree* of subintervals, discarding a subinterval I only when we discover that f takes lower values on I than on some other subinterval. (See the explanation below of how such intervals are discovered.) The result turns out to be an infinite binary tree with no infinite path, contradicting the weak Kőnig lemma.

Finding intervals to discard. If f takes lower values on a subinterval I than on some other subinterval, this is revealed by the representation of f as a sequence of pairs of rational intervals $\langle (c_n, d_n), (a_n, b_n) \rangle$ such that $f((c_n, d_n)) \subseteq (a_n, b_n)$ (see section 7.1). It follows that if I is sufficiently small then $I \subset$ some (c_m, d_m), so by enumerating the pairs in f we will eventually discover $I \subset (c_m, d_m)$ and hence that $a_m < f(x) < b_m$ for all $x \in I$. If f takes values $\geq b_m$ on some interval J we will eventually know this too, by finding $J \subset$ some (c_n, d_n) with $f((c_n, d_n)) \subseteq (a_n, b_n)$ and $a_n \geq b_m$. At this stage we know that f takes lower values on I than it does on J.

Weak Kőnig lemma implies boundedness. *If there is an unbounded continuous function on* $[0,1]$, *then there is an infinite binary tree with no infinite path.*

Proof. The intervals obtained by repeated bisection of [0,1] will be viewed as vertices of the complete binary tree: [0,1] is the top vertex, [0,1/2] and [1/2,1] are the two vertices below it, and so on. Figure 7.5 shows what the complete binary tree looks like.

We compute a subtree T of the complete binary tree in stages, at stage n enumerating the first n pairs in f and inspecting the subintervals I at level n (that is, the subintervals of length 2^{-n}). We *omit* a subinterval I if it lies below a subinterval previously omitted (to ensure that T is a tree), or if we discover that the values of f on I are less than its values on some other interval (by the process described above).

It follows that T is computable from f: to decide whether an interval I on level n belongs to T we run the above computation up to stage n. Thus the tree T exists by recursive comprehension.

Also, since f is unbounded there will be intervals *not* omitted at every level, so T is infinite. However, if T has an infinite path, this path represents a nested sequence of closed intervals with a single common

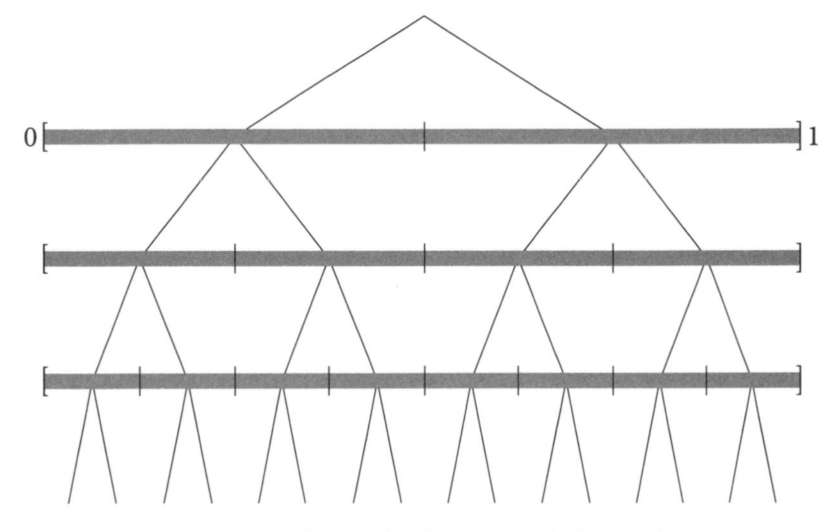

Figure 7.5 : The complete binary tree of subintervals

point, c say. And the values of f on these intervals must be unbounded—otherwise the sequence of intervals would be terminated when f was discovered to have higher values elsewhere. But then $f(c)$ is undefined, contrary to the continuity of f.

This contradiction shows that T has no infinite path.

Thus a continuous unbounded f on [0,1] implies that there is an infinite binary tree with no infinite path. Equivalently, if each infinite binary tree has an infinite path (the weak Kőnig lemma), then every continuous f is bounded. □

The above theorem is not actually needed for our proof of the extreme value theorem, but it is good motivation for it. Now, instead of an unbounded continuous function f we have one with no maximum, so instead of seeking values of $f(x)$ beyond all bounds, we seek values of $f(x)$ greater than any previously found. However, it remains true that we can discover when f takes lower values on some interval I than on some other interval, so we can apply the same construction.

Weak Kőnig lemma implies the extreme value theorem. *If there is a continuous function on* $[0,1]$ *with no maximum value, then there is an infinite binary tree with no infinite path.*

Proof. Suppose that f is a continuous function on [0,1] that takes no maximum value. As above, we repeatedly bisect the interval [0,1], and compute from f a subtree T of the complete binary tree of subintervals by the same algorithm.

Since f is continuous and with no maximum value, for any $f(x_1)$ there is a greater $f(x_2)$, hence for sufficiently small intervals I_1 containing x_1 and I_2 we will discover that f takes smaller values on I_1 than on I_2, and hence we will omit the interval I_1 from T. As in the previous proof, an interval is omitted only when some other interval is retained, so T is infinite.

Again, as in the previous proof, an infinite path in T represents a nested sequence of intervals with a single common point c. And the values of f on these intervals must include values at least as great as any value of f found elsewhere, otherwise the sequence would be terminated when a greater value of f was discovered. But then $f(c)$ exists, by continuity of f, and $f(c)$ must be the maximum value of f on [0,1].

This contradiction shows that T has no infinite path, so a continuous f on [0,1] with no maximum value implies that there is an infinite binary tree with no infinite path. Equivalently, if each infinite binary tree has an infinite path (the weak Kőnig lemma) then any continuous f on [0,1] takes a maximum value. □

7.9 THEOREMS OF WKL_0

We now define $\mathrm{WKL}_0 = \mathrm{RCA}_0$ + weak Kőnig lemma, so WKL_0 is RCA_0 with the weak Kőnig lemma as an additional set existence axiom (in fact, the weak Kőnig lemma implies recursive comprehension, so it is *the* set existence axiom for WKL_0). Then it follows from the results of the last three sections that the Heine-Borel, uniform continuity, and extreme value theorems are theorems of WKL_0. They are not theorems of RCA_0 because the weak Kőnig lemma fails in RCA_0, thanks to the counterexample in section 4.5.

WKL_0 is noteworthy because it can prove not only basic theorems such as Heine-Borel, but also some important theorems generally thought to be more difficult than Heine-Borel. Among them are the Brouwer fixed point theorem in two or more dimensions,[2] and the Jordan curve theorem.

[2]The Brouwer fixed point theorem in one dimension follows easily from the intermediate value theorem, so it is provable in RCA_0. Thus the difference between RCA_0 and WKL_0 reflects the difference between the Brouwer fixed point theorem in one and higher dimensions.

A proof of the Brouwer fixed point theorem is fairly straightforward, because one of the standard proofs—from the so-called *Sperner's lemma* of Sperner (1928)—translates into a proof in WKL_0. Much less obvious is the converse, though an outline of it may be found in Shioji and Tanaka (1990). Thus in fact the Brouwer fixed point theorem is equivalent to the weak Kőnig lemma over RCA_0.

The situation is reversed for the Jordan curve theorem. The known proof of the theorem in WKL_0—by Sakamoto and Yokoyama (2007)—is difficult (as is the classical proof), but proving that the Jordan curve theorem implies the weak Kőnig lemma is quite simple. One takes the Jordan curve theorem to say that ($\mathbb{R}^2 -$ the image of the curve) consists of two components, where points P, Q belong to the same component if there is a polygonal path from P to Q not meeting the curve. A counterexample may then be constructed using a positive continuous function f on [0,1] with no maximum (implied as above by a "failed binary tree" that is infinite but has no infinite path). From f we get the positive continuous function $1/f$ with no minimum but with greatest lower bound 0 on [0,1].

By connecting the endpoints $\langle 0, 1/f(0)\rangle$ and $\langle 1, 1/f(1)\rangle$ of the graph of $1/f$ to the endpoints of [0,1] by vertical segments one obtains a simple closed curve (suggested by figure 7.6) whose complement consists of more than two components. To see why, consider two points P and Q below the graph, but above the x-axis, and separated by x-values where $1/f(x)$ becomes arbitrarily small. Neither P nor Q is in the "outer" component of the curve complement, but they cannot be connected by a polygonal path not meeting the curve, since any such path has a minimum positive distance from the x-axis.

Sakamoto and Yokoyama (2007) also prove that a generalization of the Jordan curve theorem, the *Schönflies theorem*, is equivalent to the weak Kőnig lemma over RCA_0. The Schönflies theorem says that the interior of a simple closed curve in the plane can be mapped to a disk by a continuous bijection. A stronger version, the *Riemann mapping theorem*, implies that the interior is *conformally* equivalent to the disk: there is a continuous bijection between them that preserves angles. Horihata and Yokoyama (2014) proved the Riemann mapping theorem equivalent to arithmetical comprehension over WKL_0. This means that the Riemann mapping theorem is provable in ACA_0, but not in WKL_0, for reasons we will see in the next section.

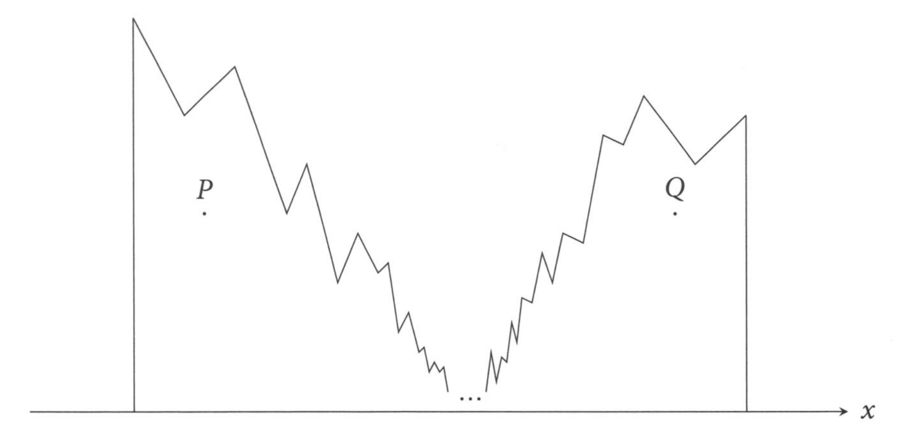

Figure 7.6 : Countering the Jordan curve theorem

Other Theorems of Topology

Brouwer (1912a) proved two other famous theorems of topology—*invariance of domain* and *invariance of dimension*—that are consequences of his fixed point theorem, proved in Brouwer (1912b). (The lemma of Sperner (1928) was in fact used by Sperner not only to prove the fixed point theorem but also to streamline the proofs of Brouwer's invariance theorems.) Invariance of domain states that a continuous injective map with open domain in $\mathbb{R}^n$ has an open image. Invariance of dimension (which follows easily from invariance of domain) states that there is no continuous bijection between $\mathbb{R}^m$ and $\mathbb{R}^n$ if $m \neq n$. The proofs of these theorems from the fixed point theorem can be carried out in RCA_0, so they are theorems of WKL_0.

However, it has apparently not yet been established whether either of the invariance theorems is provable in RCA_0 (at least for dimension greater than one). Nor do we know whether either of them implies the weak Kőnig lemma, and hence is equivalent to it. Finding the exact strength of the Brouwer invariance theorems seems to me one of the most interesting open problems in reverse mathematics.

The special case of invariance of dimension where $m = 1$ and $n = 2$ is provable in RCA_0. To prove it we suppose, for the sake of contradiction, that $f : \mathbb{R} \to \mathbb{R}^2$ is a continuous bijection. Now $\mathbb{R}$ can be *separated* by the single point $x = 0$, in the sense that there is no continuous "path" in $\mathbb{R}$ from -1 to 1 avoiding 0; that is, no continuous $t : [0,1] \to \mathbb{R} - \{0\}$ with

$t(0) = -1$ and $t(1) = 1$. Such a function t, since it does not take the value 0, contradicts the intermediate value theorem. On the other hand, there obviously *is* a continuous path in $\mathbb{R}^2$ from $f(-1)$ to $f(1)$ and avoiding $f(0)$, since these points are distinct by the assumption that f is a bijection. This contradicts the assumption that f is a continuous bijection—meaning that f^{-1} exists and is continuous—since f^{-1} would then give a continuous path in $\mathbb{R}$ from -1 to 1 avoiding 0.

This special case is provable in RCA_0 because the intermediate value theorem is. But the special case is generally considered much easier than the case of arbitrary m and n, because "separation" is a trickier concept in dimension greater than one. So provability of general invariance of dimension in RCA_0 seems doubtful. The same holds *a fortiori* for invariance of domain in higher dimensions, since invariance of domain implies invariance of dimension.

7.10 WKL_0, ACA_0, AND BEYOND

Since ACA_0 proves the weak Kőnig lemma, all theorems of WKL_0 are provable in ACA_0. In fact ACA_0 can prove more theorems than WKL_0, though the known proofs of this fact depend on some advanced results of logic and computability theory. The approach from logic will be discussed further in section 8.3, and the approach from computability theory in section 8.4. However, it is worth mentioning both briefly here.

ACA_0 can prove the *consistency* of WKL_0 by defining a class of sets that satisfy the axioms of WKL_0. But a famous theorem of logic—*Gödel's second incompleteness theorem*—implies that WKL_0 cannot prove its own consistency. So a formal statement of the consistency of WKL_0, $\mathrm{Con}(WKL_0)$, is a theorem of ACA_0 that is not a theorem of WKL_0. It follows that any theorem of ACA_0 equivalent to arithmetical comprehension, such as the full Kőnig lemma or the Bolzano-Weierstrass theorem, is not a theorem of WKL_0—otherwise WKL_0 would prove as much as ACA_0.

Another way to see this result is to show that WKL_0 can be modeled by a class of sets that does not include all arithmetically definable sets. Since any model of ACA_0 includes all arithmetically definable sets, as we saw in section 6.1, it follows that WKL_0 does not include ACA_0.

WKL_0 can in fact be modeled by a class of sets in $\Sigma_2^0 \cap \Pi_2^0$. This is shown with the help of a concept from computability theory called *low degree*. The sets of low degree are in $\Sigma_2^0 \cap \Pi_2^0$ and they have the property that any infinite tree of low degree has an infinite path of low degree. It follows that a collection of sets of low degree provides a model of WKL_0. And this model does not include all arithmetically definable sets, since not all arithmetically definable sets are in $\Sigma_2^0 \cap \Pi_2^0$, as we saw in section 2.9. For more about the concept of low degree see section 8.4, and for the key theorem about low degrees see Simpson (2009), pp. 318–319.

The "Big Five" Systems

The theorems of analysis and topology we have mentioned until now fit very neatly into the systems RCA_0, WKL_0, and ACA_0. If RCA_0 does not prove them outright then it proves them equivalent either to the weak Kőnig lemma or to arithmetical comprehension, the defining axioms of WKL_0 and ACA_0 respectively. It is therefore tempting to end the book here, with the following neat classification of theorems.

RCA_0 proves Intermediate Value Theorem.

WKL_0 proves Sequential Heine-Borel Theorem
$\Leftrightarrow$ Uniform Continuity Theorem
$\Leftrightarrow$ Extreme Value Theorem
$\Leftrightarrow$ Riemann Integrability of Continuous Functions
$\Leftrightarrow$ Brouwer Fixed Point Theorem
$\Leftrightarrow$ Jordan Curve Theorem
(Equivalences provable in RCA_0).

ACA_0 proves Kőnig Infinity Lemma
$\Leftrightarrow$ Sequential Bolzano-Weierstrass Theorem
$\Leftrightarrow$ Sequential Least Upper Bound Property
$\Leftrightarrow$ Cauchy Convergence Criterion
(Equivalences provable in RCA_0).

But mathematics never ends neatly, and in fact there are a few anomalous theorems, some of which are accommodated by two new systems, and others which seem to break the bounds of the neat classification. One such theorem is the infinite Ramsey theorem $\forall k \mathrm{RT}(k)$, mentioned in section 6.6. As we said there, the infinite Ramsey theorem is not provable in ACA_0, though some of its special cases are equivalent to arithmetical comprehension.

Thus we would hope that the infinite Ramsey theorem is provable from a stronger set existence axiom. Two such axioms have been singled out as being natural and having a variety of interesting equivalents. The weaker of the two is called *arithmetical transfinite recursion*, and adding it to PA gives a system called ATR_0. ATR_0 is strong enough to prove the infinite Ramsey theorem, in fact ATR_0 is *more* than enough, because the infinite Ramsey theorem does not imply arithmetical transfinite recursion. However, arithmetical transfinite recursion is considered a better axiom to add to PA, because it has more interesting equivalents.

The strongest set existence axiom currently considered natural is called Π^1_1 *comprehension*. The symbol Π^1_1 indicates that the properties $\varphi(n)$ defining sets are allowed to include one universal set quantifier. Such a quantifier arises in the definition of a *well-ordering* R of $\mathbb{N}$, where one says that R is a linear ordering and that every subset of $\mathbb{N}$ has a least member under the ordering relation R. The system with Π^1_1 comprehension as its set existence axiom is called Π^1_1-CA_0. The systems RCA_0, WKL_0, ACA_0, ATR_0, and Π^1_1-CA_0 are known as the "big five" because together they cover most theorems that arise in ordinary mathematics and their set existence axioms attract large families of theorems into their "orbits" (logical equivalence classes relative to the axioms of RCA_0).

The systems ATR_0 and Π^1_1-CA_0 can both handle countable well-orderings, and their theorems typically have a set-theoretic flavor. An example, equivalent to Π^1_1 comprehension, is the Cantor-Bendixson theorem stating that an uncountable closed subset of $\mathbb{R}$ is the union of a countable set and a perfect set. Theorems of this flavor are on the margins of "ordinary" analysis, so analysts may well be content with the theorems of RCA_0, WKL_0, and ACA_0. However, some important theorems of combinatorics lie above the level of ACA_0. We have already seen one—the infinite Ramsey theorem—and two even more spectacular examples are the theorems of Kruskal (1960) and Robertson and Seymour (2004).

The Theorems of Kruskal and Robertson-Seymour

Before stating the theorems in question, I would like to mention a "baby" theorem of the same type, known as the *ascending/descending sequence principle* (ADS): *any infinite sequence of rational numbers contains an infinite monotonic subsequence.* This theorem has some of the flavor of the Bolzano-Weierstrass theorem (and is a consequence of it) but it seems simpler. Nevertheless, it is not trivial because it is not provable in RCA_0. Rather mysteriously, ADS does not seem to fit neatly in the "big five" system at all: it lies somewhere below ACA_0 but is not in WKL_0. (See Hirschfeldt (2015) for the complicated known facts about ADS.)

The theorems of Kruskal and Robertson-Seymour can also be expressed in the form "infinite sequence contains an infinite monotonic subsequence," but the objects are finite graphs rather than numbers, and the orderings are relations appropriate for finite graphs rather than the usual order relation for numbers. For readers not familiar with graph theory I recommend the book Diestel (2010), which explains the order relations of "embedding" and "graph minor" and gives a proof of Kruskal's theorem. In this subsection I wish only to point out the remarkable position of the Kruskal and Robertson-Seymour theorems relative to the "big five" systems.

The usual statements of the theorems are as follows:

Kruskal's theorem. *If $T_1, T_2, T_3, \ldots$ is an infinite sequence of finite trees, then T_i embeds in T_j for some $i < j$.*

Robertson-Seymour theorem. *If $G_1, G_2, G_3, \ldots$ is an infinite sequence of finite graphs, then G_i is a minor of G_j for some $i < j$.*

We obtain equivalent statements about infinite monotonic subsequences as follows. Given an infinite sequence $T_1, T_2, T_3, \ldots$ of finite trees, consider the trees T_m that embed in *no* tree T_n with $m < n$. By Kruskal's theorem there are only finitely many such T_m. Then, if we remove these T_m from the given sequence, each remaining tree T_i embeds in some T_j with $i < j$. Therefore, we can form an infinite subsequence

$$T_{i_1} \prec T_{i_2} \prec T_{i_3} \prec \cdots,$$

where $\prec$ denotes the "embeds in" relation. A similar argument applies to finite graphs under the graph minor relation, so we can also state the theorems as follows (imitating the statement of ADS).

Kruskal's theorem. *Any infinite sequence of finite trees contains an infinite subsequence that is increasing under the "embeds in" relation.*

Robertson-Seymour theorem. *Any infinite sequence of finite graphs contains an infinite subsequence that is increasing under the graph minor relation.*

Here now are the known results, due to Friedman et al. (1987), which show the lofty position of these theorems relative to the "big five" systems: *Kruskal's theorem is not provable in* ATR_0, and *the Robertson-Seymour theorem is not provable in* $\Pi^1_1\text{-}\mathrm{CA}_0$.

CHAPTER 8

A Bigger Picture

In this book I have tried to minimize technicalities from logic and computability theory, in order to maximize understanding of reverse mathematics by ordinary mathematicians. This has meant, however, that many ideas from logic were mentioned briefly and then dropped—perhaps leaving some readers eager to know more. (I hope so!)

This final chapter aims to pick up some of the dropped ideas and set them in a bigger picture of logic and computability theory. It is still not a detailed picture, but I hope that it is a useful sketch, and that interested readers will be able to pick up more details from the sources I suggest.

The chapter begins with a sketch of *constructive* mathematics. Originally developed by a minority of mathematicians opposed to using actual infinities, constructive mathematics contributed some useful techniques for computable mathematics in systems such as RCA_0.

This is followed by sections on the completeness of logic and the incompleteness of PA and related systems. These results reveal mathematics as an arena where theorems cannot always be proved outright, but in which all of their logical *equivalents* can be found. This creates the possibility of reverse mathematics, where one seeks equivalents that are suitable as axioms.

Next we explain how computability theory helps us to distinguish the equivalence classes of theorems, and finally make a few speculative remarks on the ordering of the equivalence classes, and how this throws light on the concept of mathematical *depth*.

8.1 CONSTRUCTIVE MATHEMATICS

The discovery that real numbers correspond to sets of natural numbers was the key to the arithmetization of analysis described in chapter 2. At the same time, Cantor's 1874 discovery that $\mathbb{R}$ is uncountable raised the old specter of actual infinity. Most mathematicians overcame their fear of actual infinity because $\mathbb{R}$ was by then necessary (or at least convenient) in their work. But a few eminent mathematicians rejected actual infinity and declared that they would accept only *constructive* objects in mathematics—objects that would later be called "computable."

The first advocate of constructive mathematics was the number theorist Leopold Kronecker (1823–1891). Kronecker was in favor of arithmetization, but only as far as one could take it by constructive processes. He famously rejected the fundamental theorem of algebra (FTA) in favor of what he called his "fundamental theorem of general arithmetic," because the classical FTA located the solutions of equations in $\mathbb{C}$—a nonconstructible set. Instead he constructed, for each irreducible polynomial $p(x)$ with integer coefficients, the domain of "integer polynomials mod $p(x)$," in which the equation $p(x) = 0$ has a solution in the form of the equivalence class of x. Kronecker's construction is computable, in the modern sense, and it can be used to prove the classical FTA in RCA_0. The latter proof, perhaps, might have overcome Kronecker's objections to the classical FTA.

But later constructivists had other objections to classical mathematics—and to classical logic. The next eminent constructivist was L. E. J. Brouwer (1881–1966). Brouwer made his name by outstanding contributions to topology, such as the invariance of dimension (no continuous bijection between $\mathbb{R}^m$ and $\mathbb{R}^n$ when $m \neq n$) and the Brouwer fixed point theorem mentioned in section 7.9. Like Kronecker, Brouwer insisted that mathematical objects must be "constructed," so an object cannot be claimed to "exist" until a construction of it is given. Because of this, Brouwer rejected many classical theorems in which existence is proved without giving a construction. He likewise insisted that an object cannot be declared "nonexistent" until its existence is constructively shown to lead to contradiction. In this sense, we cannot (yet) claim that a string of 100 consecutive zeros in the decimal expansion of π is either existent or nonexistent.

For reasons like this, Brouwer rejected a law of classical logic: the *law of excluded middle*, which says that, for any proposition φ, either φ or $\neg\varphi$ is true.

The ideas of Kronecker and Brouwer are considered extreme by most mathematicians. Nevertheless, they have been useful in fields of mathematics involving computation, since the concept of "computable" seems to capture the previously vague concept of "constructive." Constructive proofs can often be put to use in RCA_0 when classical proofs are not constructive enough, and because of this RCA_0 seems to capture the content of "constructive analysis" pretty well. The match is not perfect, since RCA_0 uses classical logic, but the constructivist origins of many proofs in reverse mathematics are documented in Simpson (2009).

Thus, reverse mathematics has a big debt to constructive mathematics. It has, in my opinion, repaid its debt. By accepting classical logic and noncomputable functions, reverse mathematics can *explain* why theorems are provable in one system but not in another, and can thereby measure "how nonconstructive" are some of the theorems rejected by constructivists. The answer is often "not very." Brouwer rejected some of his own theorems, such as the fixed point theorem, as being nonconstructive.[1] But we now know, thanks to reverse mathematics, that the fixed point theorem is not far outside constructive mathematics, since it is constructively equivalent to the weak Kőnig lemma.

8.2 PREDICATE LOGIC

The language of logic has been mentioned, in passing, several times in this book. The language of PA and its extensions such as RCA_0 are part of a general language of *predicate logic*, which includes symbols for

variables: $x, y, x, \ldots, X, Y, Z, \ldots$
constants: $a, b, c, \ldots$
function symbols: $f, g, h, \ldots$
predicate symbols: $P, Q, R, \ldots$
logic symbols: $\wedge, \vee, \neg, \Rightarrow, \Leftrightarrow, \forall, \exists$, plus parentheses and commas.

[1] In 1927 Brouwer gave a lecture series in Berlin, in which he rejected the intermediate and extreme value theorems, the Bolzano-Weierstrass theorem, and finally his own fixed point theorem. See the biography of Brouwer by van Dalen (2013), pp. 382 and 503. Van Dalen's book is an excellent introduction to Brouwer's ideas in both topology and the foundations of mathematics, as well as being a fascinating account of Brouwer's life.

We have also mentioned, in passing, certain formulas that are *logically valid*; that is, true under all interpretations of the non-logic symbols. For example, for any formulas φ and ψ the following are valid:

$$\neg(\varphi \wedge \psi) \Leftrightarrow (\neg\varphi) \vee (\neg\psi),$$
$$\neg(\varphi \vee \psi) \Leftrightarrow (\neg\varphi) \wedge (\neg\psi).$$

But we have not given rules of inference for finding logically valid formulas, or even said whether a complete set of rules exists. In fact, a complete set of rules was first given by Frege (1879), though his rules were not known to be complete at the time.

The existence of a complete set of rules for generating the valid formulas was first proved by Gödel (1930). I will not give a completeness proof here, because one may be found in many textbooks of mathematical logic, and also in Stillwell (2010). But in fact a rather similar proof has already been given in section 6.7 of this book. There we showed that *any consistent set of sentences* (of PA, but the proof works for any sentences of predicate logic) *has an interpretation that makes them all true*.

By a similar argument one can show that there is a set of *falsification rules*, which falsify any sentence that is false under some interpretation. The rules repeatedly reduce the length of the sentence. For example, to falsify $\neg(\varphi \vee \psi)$ it suffices to falsify one of the shorter sentences $\neg\varphi$ or $\neg\psi$ because $\neg(\varphi \vee \psi) \Leftrightarrow (\neg\varphi) \wedge (\neg\psi)$. The rules succeed by breaking formulas down to their smallest parts, at which point it is clear whether they can be falsified or not. Now, the reverse of a falsification rule is a *rule for proof*—in this case: from $\neg\varphi$ or $\neg\psi$ infer $\neg(\varphi \vee \psi)$. In this way, the completeness of the falsification process gives a complete set of rules for proving all logically valid formulas.

This argument, I think, gives the general idea of the completeness proof, and how it is related to the theorem in section 6.7. What the two have in common is reliance on the weak Kőnig lemma to obtain an infinite path in a tree: in section 6.7 to find an interpretation satisfying a consistent set of formulas; here to find an interpretation falsifying an invalid formula. This is no accident. *The completeness theorem for predicate logic is yet another equivalent of the weak Kőnig lemma.*

This equivalence reveals some noncomputability in predicate logic. Indeed it was proved by Church (1936a) and Turing (1936) that *the problem of deciding validity in predicate logic is unsolvable*. That is, there is no algorithm for deciding, given an arbitrary formula φ of predicate logic, whether φ is logically valid. Once unsolvable problems in computation are known the unsolvability of the validity problem is not a great surprise. It was already known in 1936 that computation can be arithmetized, and this amounts to translating computation into deductions, in predicate logic, from the axioms of PA. In fact, one can bypass arithmetization and translate computation into predicate logic directly. This is how Turing (1936) proved unsolvability of the validity problem, or the *Entscheidungsproblem* ("decision problem" in German) as it was then known.

At any rate, the undecidability of the Entscheidungsproblem throws new light on Gödel's completeness theorem. It means that the *valid formulas of predicate logic are computably enumerable, but the invalid formulas are not*—thus there is a sense in which truth is more accessible than falsehood! It also means that we can computably enumerate all theorems that RCA_0 proves equivalent to, say, the weak Kőnig lemma or the monotone convergence theorem, or any other statement we care to assume. This is the best we can hope for, because it is not generally possible to prove theorems outright, as we will see from many angles in the next section.

The main value of an axiom system for mathematics, such as RCA_0, is its ability to prove *equivalences* between theorems it cannot prove outright. We can view RCA_0 as shown in figure 8.1: as a "planet" of theorems proved outright, surrounded by "rings" or "orbits" of theorems that RCA_0 can prove equivalent to each other. Thus the ring WKL_0 contains all theorems in the orbit of the weak Kőnig lemma, the ring ACA_0 contains all theorems in the orbit of the monotone convergence theorem, and there are infinitely many rings further out, because there is no end to the theorems a consistent system cannot prove.

This is the message of Gödel *incompleteness*, which we study in the next section. Gödel completeness mitigates the effect of Gödel incompleteness, a little, by allowing all equivalents of a given unprovable sentence to be found.

The hard part of axiomatic mathematics is showing unprovability of sentences in the base theory. It has always been so. As we saw in chapter 1, it was hard to show that the parallel axiom is not provable from Euclid's other axioms, and hard to show that the axiom of choice is not provable in ZF. In both cases great ingenuity was needed to construct a model of the base theory in which the axiom in question did not hold. The same

Figure 8.1 : RCA_0 and two of its rings. (From a picture of Saturn and its rings created from images obtained by NASA's Cassini spacecraft on Oct. 10, 2013, by Gordan Ugarkovic.)

is true, though perhaps not to such an epic extent, of the axioms not provable in RCA_0, such as the weak and strong Kőnig lemmas. We say more about our evolving understanding of unprovability in the next two sections.

8.3 VARIETIES OF INCOMPLETENESS

Section 4.6 outlined a connection between unsolvability and incompleteness, and sections 4.7 and 7.10 mentioned the unprovability of consistency. It is worth saying a little more about different versions of incompleteness, because the step from each kind to the next involves an interesting idea.

1. The computably enumerable, noncomputable set D gives the existence of unprovable theorems of the form $n \notin D$ in any formal system.
   As we said in section 4.6, D is not computable because its complement $\mathbb{N} - D$ is not computably enumerable, being different from the nth computably enumerable set (which we will now call W_n) with respect to the number n. Namely: $n \in D \Leftrightarrow n \in W_n$, so $n \in \mathbb{N} - D \Leftrightarrow n \notin W_n$.
   But a formal system, by definition, has a computably enumerable set of theorems, so we can computably enumerate its theorems of the form $n \notin D$. Since $\mathbb{N} - D$ is *not* computably enumerable, these theorems (if correct) do not include all true statements of the form $n \notin D$.
2. Suppose that our formal system $\mathcal{F}$ is consistent and that the set of n for which $\mathcal{F}$ proves "$n \notin W_n$" is the computably enumerable set W_m. We can also assume that $\mathcal{F}$ is strong enough to prove $n \in W_n$ whenever this is true, because there is a computable enumeration of all the W_n. What can we say about the sentence "$m \notin W_m$"?
   If $\mathcal{F}$ proves "$m \notin W_m$," then $m \in W_m$, by definition of W_m. But we have assumed that $\mathcal{F}$ can prove all such true statements, so $\mathcal{F}$ also proves "$m \in W_m$," contradicting its consistency. So $\mathcal{F}$ does *not* prove "$m \notin W_m$," which means $m \notin W_m$, by the definition of W_m again. Thus "$m \notin W_m$" is a *specific true sentence* that $\mathcal{F}$ fails to prove.
   Notice that the sentence "$m \notin W_m$" essentially says "I am not provable," because $m \notin W_m$ means "$m \notin W_m$" is not provable, by definition of W_m.

3. By arithmetizing computation we can find a specific unprovable sentence of PA that is equivalent to $m \notin W_m$, and hence is unprovable, *assuming that* PA *is consistent*. Arithmetization of formal deduction in PA enables us to express the consistency of PA by a sentence in the language of PA, Con(PA).
4. Version 2 shows that "Con(PA) $\Rightarrow m \notin W_m$," and version 3 shows that the proof can be carried out in PA. It follows that Con(PA) cannot be proved in PA, otherwise modus ponens would give a proof of "$m \notin W_m$," and we know that "$m \notin W_m$" has no proof in PA. Thus PA, if consistent, *cannot prove its own consistency*, and neither can any consistent system that includes PA, because a similar argument would apply.[2]

This is a brilliant train of thought—one of the most stunning in mathematics. Versions 1 and 2 are like those found by Post in the 1920s, and reviewed by him in Post (1944). Versions 3 and 4 are essentially the first and second incompleteness theorems of Gödel (1931), though Gödel found the first theorem differently, by directly constructing a sentence that says "I am not provable." Interestingly, Gödel first proved incompleteness for a higher-level system of mathematics because he did not immediately realize that computation could be arithmetized. He was prompted to arithmetize by von Neumann, according to Wang (1981):

> In September 1930 ... Gödel announced his result ... Von Neumann was very enthusiastic about the result and had a private discussion with Gödel. In this discussion, von Neumann asked whether number-theoretical undecidable propositions could also be constructed ... and expressed his belief that it could be done. ... Shortly afterward Gödel, to his own astonishment, succeeded in turning the undecidable proposition into a polynomial form preceded by quantifiers (over natural numbers).

Von Neumann also deserves credit for the second incompleteness theorem, because he pointed out unprovability of consistency in a letter to Gödel (von Neumann (1930)) before Gödel announced the result himself. These incompleteness theorems are a wonderful feat of logic but, of

[2]Moreover, the argument applies to certain weaker systems in which computation is representable, such as RCA_0 and WKL_0. This is why we were able to say, in section 7.10, that WKL_0 cannot prove Con(WKL_0).

the unprovable sentences found so far, only Con(PA) is of intrinsic interest, and it is interesting mainly to logicians. In fact, all known unprovable sentences of PA have been devised by logicians. The one most nearly like "normal" mathematics is the Paris-Harrington theorem mentioned in section 6.6.

RCA_0 is incomplete in an interesting way because it fails to prove some ordinary mathematical sentences, such as the monotone convergence theorem. And this unprovability comes *directly* from a computably enumerable, noncomputable set, with no logical contortions. By seeking unprovability in systems other than PA, reverse mathematics finds well-known theorems not provable from (what some people think are) natural axiom systems.

8.4 COMPUTABILITY

Degrees of Unsolvability

In section 4.3 we introduced an enumeration of computable partial functions $\Phi_1, \Phi_2, \Phi_3, \ldots$ and mentioned that the computably enumerable sets are the domains of these functions. We now let W_n denote the domain of Φ_n, so $W_1, W_2, W_3, \ldots$ is an enumeration of all computably enumerable sets. Since $\Phi_n(m)$ is a computable partial function of the two variables m, n, the set $K = \{P(m,n) : m \in W_n\}$ (where P is the pairing function introduced in section 2.4) is computably enumerable and it encodes the sequence $W_1, W_2, W_3 \ldots$. We call K a *universal* computably enumerable set.

Post (1944) introduced essentially the same set and observed that the *membership problem* for any computably enumerable set is "reducible" to that of K, in the following way: to decide whether $m \in W_n$, compute $P(m,n)$ and ask whether $P(m,n) \in K$. It follows that the membership problem for K is *unsolvable* because we know from section 4.3 that there are particular W_m with unsolvable membership problem, such as the set D defined there.

More generally, we may be able to "reduce" a set $A \subseteq \mathbb{N}$ to a set $B \subseteq \mathbb{N}$ by an algorithm that correctly answers each question "Does $n \in A$?" when given the answers to all questions of the form "Does $m \in B$?" Post called this reducibility notion *Turing reducibility* because it was introduced briefly by Turing (1939). We now denote Turing reducibility by $\leq_T$, so the reducibility of D to K can be written $D \leq_T K$. It so happens (though it is not obvious) that $K \leq_T D$ too, in which case we say that D and K have the same *Turing degree*, or the same *degree of unsolvability*.

So far we know two degrees of unsolvability. All computable sets have the same (trivial) degree, denoted by $\mathbf{0}$, and somewhere above $\mathbf{0}$ is the degree of K and D, denoted by $\mathbf{0}'$ and called the *Turing jump* of $\mathbf{0}$. It can be shown, with some difficulty, that there are degrees of unsolvability between $\mathbf{0}$ and $\mathbf{0}'$, but finding degrees greater than $\mathbf{0}'$ is easy by iterating the Turing jump.

We apply the Turing jump to any set $X \subseteq \mathbb{N}$ by giving algorithms the answers to all questions "Does $m \in X$?" This yields sets $W_1^X, W_2^X, W_3^X, \ldots$ *computably enumerable in* X and one, $K^X = \{P(m,n) : m \in W_n^X\}$, with maximal Turing degree. The degree of K^X is called the Turing jump of the degree of X. In particular, when $X = K$ the degree of K^X is $\mathbf{0}''$, the double Turing jump of $\mathbf{0}$. By iterating this construction we obtain an infinite ascending sequence of degrees $\mathbf{0} <_T \mathbf{0}' <_T \mathbf{0}'' <_T \mathbf{0}''' <_T \cdots$.

And Arithmetically Definable Sets

As we know from chapter 5, the computably enumerable sets are precisely the Σ_1^0 sets. So it follows from the results of the previous subsection that each Σ_1^0 set has Turing degree $\leq_T \mathbf{0}'$. It is not the case that every set of degree $\leq_T \mathbf{0}'$ is Σ_1^0, but it can be proved that all such sets are in Σ_2^0, by arithmetizing the concept of computation relative to the Σ_1^0 set K.

More generally, we can prove that any set in Σ_n^0 has degree $\leq_T \mathbf{0}^{(n)}$ (the nth jump of $\mathbf{0}$) and that any set of degree $\leq_T \mathbf{0}^{(n)}$ is in Σ_{n+1}^0. So *the arithmetically definable sets are precisely those of degree* $\leq_T \mathbf{0}^{(n)}$ *for some* n. Also, $\mathbf{0}^{(n)}$ is the maximal degree of sets in Σ_n^0, so the "arithmetic complexity" of a set (measured by the number of quantifiers in its definition) keeps pace with its "computational complexity" (measured by the number of jumps). A classic account of the relation between computability and the arithmetically definable sets is in Rogers (1967).

Low Degrees

The parallel between arithmetic complexity and computational complexity enables us to bound the arithmetic complexity of sets by bounding their computational complexity. This is the right way to bound the complexity of models of WKL_0, and hence to show that WKL_0 is weaker than ACA_0.

The appropriate computational concept is that of *low degree*, defined to be a Turing degree whose jump is $\mathbf{0}'$. A theorem of computability theory says that *an infinite tree of low degree has an infinite path of low degree* (for a proof, see Simpson (2009), p. 318), so a collection of low-degree subsets of $\mathbb{N}$ satisfies the weak Kőnig lemma and hence is a model of WKL_0.

But this collection does *not* include all arithmetically definable sets (in fact, it includes only Σ_2^0 sets), so *it is not a model of* ACA_0. This explains more fully the claim in section 7.10 that ACA_0 is stronger than WKL_0.

In fact, both proofs mentioned in section 7.10 can be based on low degree sets. One proof is the one just described: some low degree sets form a model of WKL_0 but not of ACA_0 because they do not include all arithmetically definable sets. The other proof argues that this collection of low degree sets can be defined in ACA_0, by arithmetizing computation. Thus ACA_0 can define a model of WKL_0, and hence prove the consistency of WKL_0. WKL_0 itself cannot do this, by Gödel's second incompleteness theorem, so ACA_0 is stronger than WKL_0.

8.5 SET THEORY

Beyond the basic analysis discussed in this book there are areas that rely on set theory, particularly the axiom of choice. An example is measure theory. A certain amount of measure theory can be done in weak systems such as WKL_0, in fact in a system called WWKL_0 with an even weaker Kőnig lemma. But none of the "big five" can touch what is arguably the fundamental question of measure theory: *which subsets of* $\mathbb{R}$ *are measurable?* The answer to this question depends on AC, and more.

For example, AC implies that nonmeasurable sets exist, but the nature of the nonmeasurable sets depends on further axioms not provable in ZF. In ZF alone, it is consistent for all subsets of $\mathbb{R}$ to be measurable, while weaker choice axioms still hold. For an introduction to the complex interplay between measure theory and set theory, see Stillwell (2013).

Axioms of choice also control the existence of objects in algebra. Structures such as rings, fields, and vector spaces are virtually as "arbitrary" as arbitrary sets, so associated objects such as

- maximal ideal of a ring,
- algebraic closure of a field, and
- basis of a vector space

can generally not be explicitly defined, only "called into being" by some axiom of choice. Algebraists frequently use a general-purpose equivalent of AC known as *Zorn's lemma*, which states the existence of a maximal object in a collection of sets partially ordered by containment.

For example, among the collection of independent sets in a vector space, a maximal element is a basis. We have already mentioned, in section 1.5, that existence of a basis is equivalent to AC. The existence of maximal ideals and algebraic closures is also not provable in ZF, though choice axioms weaker than AC suffice to prove them. Proofs from Zorn's lemma of the three results above may be found, for example, in Abhyankar (2006), pp. 46–50.

If we consider only *countable* algebraic structures, then they are well-ordered by definition and the need for AC disappears. Instead, by coding the structures by sets of natural numbers, reverse mathematics can come into play, and indeed the existence of maximal structures is nicely related to RCA_0, WKL_0, and ACA_0:

- Existence of a maximal ideal in a countable commutative ring is equivalent to ACA_0.
- Existence of the algebraic closure of countable field is provable in RCA_0.
- Uniqueness of the algebraic closure of a countable field is equivalent to WKL_0.
- Existence of a basis for a countable vector space over $\mathbb{Q}$ is equivalent to ACA_0.

These theorems are due to Friedman et al. (1983), and proofs may be found in Simpson (2009). They reveal quite surprising equivalences, over RCA_0, between theorems in analysis and countable algebra. To take a random example, the Cauchy convergence criterion is equivalent to the existence of a maximal ideal in a countable ring.

AC in Elementary Analysis

As mentioned above, ZF alone cannot decide certain questions of advanced analysis, such as the existence of nonmeasurable sets of real numbers. In elementary analysis it is possible to avoid using the full strength of ZF—indeed we have seen that ACA_0 is strong enough to prove the basic theorems, and even quite advanced ones such as the Riemann mapping theorem. But to use ACA_0 we have to formulate the basic theorems appropriately. In particular, we have to avoid mention of arbitrary sets of real numbers or intervals in theorems such as Bolzano-Weierstrass or Heine-Borel.

The strongest formulation of Bolzano-Weierstrass, which appears in some textbooks, is the following: *a bounded infinite set of real numbers contains a convergent subsequence.* This formulation of Bolzano-Weierstrass is not even provable in ZF, because in ZF it is not possible to prove that an infinite set S of real numbers contains an infinite sequence $s_1, s_2, s_3, \ldots$. The obvious way to prove this claim, of course, is to choose a member $s_1 \in S$, then a member $s_2 \in S - \{s_1\}$, then a member $s_3 \in S - \{s_1, s_2\}$, and so on. Since S is infinite, this process gives an infinite sequence $s_1, s_2, s_3, \ldots$ contained in S *but* it involves an infinite sequence of *choices*, and hence appeals to AC. Cohen (1966) in fact constructed a model of ZF in which there is a bounded infinite set of real numbers containing *no* infinite sequence, so *the strong formulation of Bolzano-Weierstrass is not provable in* ZF.

The same model of ZF can be used to construct a function f which is "sequentially continuous" at a certain point $x = a$—in the sense that $f(a_n)$ has limit $f(a)$ for each sequence a_n with limit a—although f is not continuous at $x = a$. Under AC, sequential continuity at a point is equivalent to actual continuity at a point. So *this equivalence is not provable in* ZF, although it is assumed in some textbooks of analysis, such as Abbott (2015).

8.6 CONCEPTS OF "DEPTH"

Mathematicians often call a theorem "deep" if it is some way fundamental, revealing, and hard to prove. The proof of a deep theorem is often uncovered only through the work of generations of mathematicians, by which time the theorem has been found to underlie many others. Examples from the recent history of mathematics are the prime number theorem, the classification of finite simple groups, and the graph minor theorem mentioned in section 7.10.

The deepest theorems of mathematics are understood by so few people that there is probably no hope of saying precisely what makes them deep. A more realistic goal is to look at the "relative depth" of some more approachable theorems. Under what conditions might we say that Theorem A is "deeper than" Theorem B?

One way, which should suggest itself to any reader of this book, is to find a natural axiom system that proves Theorem B but not Theorem A. With this criterion, the axiom systems considered in this book allow us to rank many theorems by relative depth. For example:

- The parallel axiom is deeper than the theorems proved by Euclid's other axioms.
- AC is deeper than the theorems proved by ZF.
- The extreme value theorem is deeper than the intermediate value theorem.
- The sequential Bolzano-Weierstrass theorem is deeper than the sequential Heine-Borel theorem.
- The Riemann mapping theorem is deeper than the Jordan curve theorem.

In all of these examples, we find theorems that are not only harder to prove, but also *fundamental*, in the sense of underlying many other theorems. Thus reverse mathematics has exposed a precise and new mathematical concept, which might well be an indicator of "depth."

However, we must admit that reverse mathematics has so far revealed signs of "depth" only in theorems about infinite objects, such as real numbers or subsets of $\mathbb{N}$. There has been little progress on theorems about the natural numbers. The only examples where we might say Theorem A is "deeper than" Theorem B is where Theorem B is provable in PA and Theorem A is not. Few such examples are known (see section 7.10), and they do not include theorems that number theorists really think are deep, such as the prime number theorem.

As mentioned in section 6.8, there was once thought to be a distinction between "elementary" methods in number theory and methods that essentially involve analysis. That idea evaporated when an elementary proof of the prime number theorem was found in 1949, and in section 6.8 we have seen how reverse mathematics confirms that analysis is inessential in number theory—at least the analysis available in ACA_0.

Still, one feels there ought to be a distinction between "elementary" and "analytic" methods. Finding it remains a challenge for the reverse mathematics of the future.

Bibliography

Abbott, S. (2015). *Understanding Analysis* (Second ed.). Undergraduate Texts in Mathematics. Springer, New York.

Abhyankar, S. S. (2006). *Lectures on Algebra. Vol. I.* World Scientific Publishing Co. Pte. Ltd., Hackensack, NJ.

Beltrami, E. (1868). Teoria fondamentale degli spazii di curvatura costante. *Annali di Matematica Pura ed Applicata, ser. 2, 2*, 232–255. In his *Opere Matematiche* 1: 406–429, English translation in Stillwell (1996).

Blass, A. (1984). Existence of bases implies the axiom of choice. In *Axiomatic Set Theory (Boulder, Colorado, 1983)*, Volume 31 of *Contemporary Mathematics*, pp. 31–33. American Mathematical Society, Providence, RI.

Bolyai, F. (1832). *Tentamen juventutem studiosam in elementa matheseos purae, elementaris ac sublimioris, methodo intuitiva, evidentiaque huic propria, introducendi.* Marosvásárhely.

Bolzano, B. (1817). *Rein analytischer Beweis des Lehrsatzes dass zwischen je zwey Werthen, die ein entgegengesetzes Resultat gewähren, wenigstens eine reelle Wurzel der Gleichung liege.* Ostwald's Klassiker, vol. 153. Engelmann, Leipzig, 1905. English translation in Russ (2004), 251–277.

Borel, É. (1898). *Leçons sur la théorie des fonctions.* Paris: Gauthier-Villars.

Bradley, R. E. and C. E. Sandifer (2009). *Cauchy's Cours d'analyse.* Sources and Studies in the History of Mathematics and Physical Sciences. Springer, New York.

Brouwer, L. E. J. (1912a). Beweis der Invarianz des n-dimensionalen Gebiets. *Mathematische Annalen 71*, 305–315.

Brouwer, L. E. J. (1912b). Über Abbildungen von Mannigfaltigkeiten. *Mathematische Annalen 71*, 97–115.

Cantor, G. (1874). Über eine Eigenschaft des Inbegriffes aller reellen algebraischen Zahlen. *Journal für reine und angewandte Mathematik 77*, 258–262. In his *Gesammelte Abhandlungen*, 145–148. English translation by W. Ewald in Ewald (1996), Vol. II, 840–843.

Cantor, G. (1883). Über unendliche, lineare Punktmannigfaltigkeiten. *Mathematische Annalen 21*, 545–591. English translation by William Ewald in Ewald (1996), volume II, pp. 881–920.

Cantor, G. (1891). Über eine elementare Frage der Mannigfaltigkeitslehre. *Jahresbericht deutschen Mathematiker-Vereinigung 1*, 75–78. English translation by W. Ewald in Ewald (1996), Vol. II, 920–922.

Cantor, G. (1895). Beiträge zur Begründung der transfiniten Mengenlehre. *Mathematische Annalen 46*(4), 481–512. English translation by P. E. B. Jourdain in Cantor (1952).

Cantor, G. (1952). *Contributions to the founding of the theory of transfinite numbers.* Dover

Publications, Inc., New York, N. Y. Translated, and provided with an introduction and notes, by Philip E. B. Jourdain.

Cauchy, A.-L. (1821). *Cours d'Analyse*. Chez Debure Frères. Annotated English translation in Bradley and Sandifer (2009).

Church, A. (1936a). A note on the Entscheidungsproblem. *Journal of Symbolic Logic 1*, 40–41.

Church, A. (1936b). An unsolvable problem in elementary number theory. *American Journal of Mathematics 58*, 345–363.

Cohen, P. (1963). The independence of the continuum hypothesis I, II. *Proceedings of the National Academy of Sciences 50, 51*, 1143–1148, 105–110.

Cohen, P. J. (1966). *Set Theory and the Continuum Hypothesis*. W. A. Benjamin, Inc., New York-Amsterdam.

Davis, M. (Ed.) (2004). *The Undecidable*. Mineola, NY: Dover Publications Inc. Corrected reprint of the 1965 original [Raven Press, Hewlett, NY].

Dawson, Jr., J. W. (2015). *Why Prove it Again?* Springer, Cham. Alternative proofs in mathematical practice. With the assistance of Bruce S. Babcock and with a chapter by Steven H. Weintraub.

Dedekind, R. (1872). *Stetigkeit und irrationale Zahlen*. Braunschweig: Vieweg und Sohn. English translation in: *Essays on the Theory of Numbers*, Dover, New York, 1963.

Descartes, R. (1637). *The geometry of René Descartes. (With a facsimile of the first edition, 1637.)*. New York, NY: Dover Publications Inc. Translated by David Eugene Smith and Marcia L. Latham, 1954.

Diestel, R. (2010). *Graph Theory* (Fourth ed.), Volume 173 of *Graduate Texts in Mathematics*. Springer, Heidelberg.

Ellerman, D. (2014). On Double-Entry Bookkeeping: The Mathematical Treatment. *Accounting Education 23*(5), 483–501.

Erdős, P. (1949). On a new method in elementary number theory which leads to an elementary proof of the prime number theorem. *Proceedings of the National Academy of Sciences U. S. A. 35*, 374–384.

Ewald, W. (1996). *From Kant to Hilbert: A Source Book in the Foundations of Mathematics. Vol. I, II*. The Clarendon Press, Oxford University Press, New York.

Fraenkel, A. (1922). Zu den Grundlagen der Cantor-Zermeloschen Mengenlehre. *Mathematische Annalen 86*, 230–237.

Frege, G. (1879). *Begriffschrift*. English translation in van Heijenoort (1967), pp. 5–82.

Friedman, H. (1975). Some systems of second order arithmetic and their use. In *Proceedings of the International Congress of Mathematicians (Vancouver, B. C., 1974), Vol. 1*, pp. 235–242. Canadian Mathematical Congress, Montreal, Quebec.

Friedman, H. (1976). Systems of second order arithmetic with restricted induction I, II. *Journal of Symbolic Logic 41*, 557–559.

Friedman, H., N. Robertson, and P. Seymour (1987). The metamathematics of the graph minor theorem. In *Logic and Combinatorics*, pp. 229–261. American Mathematical Society.

Friedman, H. M., S. G. Simpson, and R. L. Smith (1983). Countable algebra and set existence axioms. *Annals of Pure and Applied Logic 25*(2), 141–181.

Gauss, C. F. (1816). Demonstratio nova altera theorematis omnem functionem algebraicum rationalem integram unius variabilis in factores reales primi vel secundi gradus resolvi posse. *Commentationes societas regiae scientiarum Gottingensis recentiores 3*, 107–142. In his *Werke* 3: 31–56.

Gödel, K. (1930). Die Vollständigkeit der Axiome des logischen Funktionenkalküls. *Monatshefte für Mathematik und Physik 37*, 349–360.

Gödel, K. (1931). Über formal unentscheidbare Sätze der Principia Mathematica und verwandter Systeme. I. *Monatshefte für Mathematik und Physik 38*, 173–198. English translation in van Heijenoort (1967), 596–616.

Gödel, K. (2014). *Collected Works. Vol. V. Correspondence H–Z.* The Clarendon Press, Oxford University Press, Oxford. Edited by Solomon Feferman, John W. Dawson, Jr., Warren Goldfarb, Charles Parsons, and Wilfried Sieg, Paperback edition of the 2003 original.

Grassmann, H. (1844). *Die lineale Ausdehnungslehre.* Otto Wiegand, Leipzig. English translation in Grassmann (1995), pp. 1–312.

Grassmann, H. (1847). *Geometrische Analyse geknüpft an die von Leibniz gefundene Geometrische Charakteristik.* Weidmann'sche Buchhandlung, Leipzig. English translation in Grassmann (1995), pp. 313–414.

Grassmann, H. (1861). *Lehrbuch der Arithmetic.* Berlin: Enslin.

Grassmann, H. (1995). *A New Branch of Mathematics.* Open Court Publishing Co., Chicago, IL. The *Ausdehnungslehre* of 1844 and other works, Translated from the German and with a note by Lloyd C. Kannenberg. With a foreword by Albert C. Lewis.

Hamel, G. (1905). Eine Basis aller Zahlen und die unstetigen Lösungen der Funktionalgleichung $f(x + y) = f(x) + f(y)$. *Mathematische Annalen 60*, 459–462.

Harmilton, W. R. (1835). Theory of conjugate functions, or algebraic couples. Communicated to the Royal Irish Academy, 1 June 1835. In his *Mathematical Papers 76–96.*

Hardy, G. H. and H. Heilbron (1938). Edmund Landau. *Journal of the London Mathematical Society 13.*

Hausdorff, F. (1914). *Grundzüge der Mengenlehre.* Veit and Comp.

Heath, T. L. (1956). *The Thirteen Books of Euclid's Elements translated from the text of Heiberg. Vol. I: Introduction and Books I, II. Vol. II: Books III–IX. Vol. III: Books X–XIII and Appendix.* New York: Dover Publications Inc. Translated with introduction and commentary by Thomas L. Heath, 2nd ed.

Hilbert, D. (1899). *Grundlagen der Geometrie.* Leipzig: Teubner. English translation: *Foundations of Geometry.* Open Court, Chicago, 1971.

Hilbert, D. (1902). Mathematical problems. *Bulletin of the American Mathematical Society 8*, 437–479. Translated by Frances Winston Newson.

Hirschfeldt, D. R. (2015). *Slicing the Truth*, Volume 28 of *Lecture Notes Series. Institute for Mathematical Sciences. National University of Singapore.* World Scientific Publishing Co. Pte. Ltd., Hackensack, NJ. On the computable and reverse mathematics of combinatorial principles. Edited and with a foreword by Chitat Chong, Qi Feng, Theodore A. Slaman, W. Hugh Woodin, and Yue Yang.

Horihata, Y. and K. Yokoyama (2014). Nonstandard second-order arithmetic and Riemann's mapping theorem. *Annals of Pure and Applied Logic 165*(2), 520–551.

Kőnig, D. (1927). Über eine Schlussweise aus dem Endlichen ins Unendliche. *Acta Litterarum ac Scientiarum Regiae Universitatis Hungaricae Francisco-Josephinae, sectio scientiarum mathematicarum 3*, 121–130.

Kruskal, J. B. (1960). Well-quasi-ordering, the Tree Theorem, and Vazsonyi's conjecture. *Transactions of the American Mathematical Society 95*, 210–225.

Lambert, J. H. (1766). Die Theorie der Parallellinien. *Magazin für reine und angewandte Mathematik (1786)*, 137–164, 325–358.

Matijasevič, Y. V. (1971). Diophantine representation of recursively enumerable predi-

cates. In *Actes du Congrès International des Mathématiciens (Nice, 1970), Tome 1*, pp. 235–238. Gauthier-Villars, Paris.
Minkowski, H. (1908). Raum und Zeit. *Jahresbericht der Deutschen Mathematiker-Vereinigung 17*, 75–88.
Pacioli, L. (1494). *Ancient Double-Entry Bookkeeping. Lucas Pacioli's Treatise*. John B. Geijsbeek, Denver, CO, 1914. Accounting section of Pacioli's *Summa de Arithmetica* of 1494, translated by John B. Geijsbeek.
Paris, J. and L. Harrington (1977). A mathematical incompleteness in Peano arithmetic. In *Handbook of Mathematical Logic*, ed. J. Barwise, North-Holland, Amsterdam.
Peano, G. (1888). *Calcolo Geometrico secondo l'Ausdehnungslehre di H. Grassmann, preceduto dalle operazioni della logica deduttiva*. Bocca, Turin. English translation in Peano (2000).
Peano, G. (1889). *Arithmetices principia*. Torino: Bocca.
Peano, G. (2000). *Geometric Calculus*. Birkhäuser Boston, Inc., Boston, MA. According to the *Ausdehnungslehre* of H. Grassmann, Translated from the Italian by Lloyd C. Kannenberg.
Poincaré, H. (1902). Du rôle de l'intuition et de la logique en mathématiques. *Compte Rendu du Deuxième Congrès International des Mathématiciens, Paris*.
Pólya, G. and R. Fueter (1923). Rationale Abzählung der Gitterpunkte. *Vierteljahrschrifft der Naturforschende Gesellschaft in Zürich 58*.
Post, E. L. (1936). Finite combinatory processes - formulation 1. *Journal of Symbolic Logic 1*, 103–105.
Post, E. L. (1941). Absolutely unsolvable problems and relatively undecidable propositions - an account of an anticipation. In Davis (2004), pp. 338–433.
Post, E. L. (1944). Recursively enumerable sets of positive integers and their decision problems. *Bulletin of the American Mathematical Society 50*, 284–316.
Post, E. L. (1947). Recursive unsolvability of a problem of Thue. *Journal of Symbolic Logic 12*, 1–11.
Ramsey, F. P. (1930). On a problem of formal logic. *Proceedings of the London Mathematical Society 30*, 264–286.
Robertson, N. and P. D. Seymour (2004). Graph minors. XX. Wagner's conjecture. *Journal of Combinatorial Theory Series B 92*(2), 325–357.
Rogers, Jr., H. (1967). *Theory of Recursive Functions and Effective Computability*. McGraw-Hill Book Co., New York-Toronto, Ont.-London.
Russ, S. (2004). *The Mathematical Works of Bernard Bolzano*. Oxford: Oxford University Press.
Saccheri, G. (1733). *Euclid Vindicated from Every Blemish*. Classic Texts in the Sciences. Birkhäuser/Springer, Cham, 2014. Dual Latin-English text, edited and annotated by Vincenzo De Risi. Translated from the Italian by G. B. Halsted and L. Allegri.
Sakamoto, N. and K. Yokoyama (2007). The Jordan curve theorem and the Schönflies theorem in weak second-order arithmetic. *Archive for Mathematical Logic 46*(5-6), 465–480.
Selberg, A. (1949). An elementary proof of the prime-number theorem. *Annals of Mathematics. Second Series 50*, 305–313.
Shioji, N. and K. Tanaka (1990). Fixed point theory in weak second-order arithmetic. *Annals of Pure and Applied Logic 47*(2), 167–188.
Sieg, W. (2013). *Hilbert's Programs and Beyond*. Oxford University Press, Oxford.
Simpson, S. G. (2009). *Subsystems of second order arithmetic* (Second ed.). Perspectives

in Logic. Cambridge University Press, Cambridge; Association for Symbolic Logic, Poughkeepsie, NY.

Smith, H. J. S. (1875). On the integration of discontinuous functions. *Proceedings of the London Mathematical Society 6*, 140–153.

Smoryński, C. (1991). *Logical number theory. I.* Universitext. Springer-Verlag, Berlin.

Smullyan, R. M. (1961). *Theory of Formal Systems.* Annals of Mathematics Studies, No. 47. Princeton, NJ: Princeton University Press.

Sperner, E. (1928). Neuer Beweis für die Invarianz der Dimensionszahl und des Gebietes. *Abhandlungen aus dem mathematischen Seminar der Universität Hamburg 6*, 265–272.

Stillwell, J. (1996). *Sources of Hyperbolic Geometry.* Providence, RI: American Mathematical Society.

Stillwell, J. (2010). *Roads to Infinity.* Natick, MA: A K Peters Ltd.

Stillwell, J. (2013). *The Real Numbers.* Undergraduate Texts in Mathematics. Springer, Cham. An introduction to set theory and analysis.

Thue, A. (1914). Probleme über Veränderungen von Zeichenreihen nach gegebenen Regeln. J. Dybvad, Kristiania, 34 pages.

Turing, A. (1936). On computable numbers, with an application to the Entscheidungsproblem. *Proceedings of the London Mathematical Society 42*, 230–265.

Turing, A. M. (1939). Systems of Logic Based on Ordinals. *Proceedings of the London Mathematical Society 45*(1), 161–228.

Turing, A. M. (1950). The word problem in semi-groups with cancellation. *Annals of Mathematics (2) 52*, 491–505.

van Dalen, D. (2013). *L. E. J. Brouwer—Topologist, Intuitionist, Philosopher.* Springer, London.

van Heijenoort, J. (1967). *From Frege to Gödel. A Source Book in Mathematical Logic, 1879–1931.* Cambridge, MA: Harvard University Press.

von Neumann, J. (1930). Letter to Gödel, 20 November 1930, in Gödel (2014), p. 337.

Wang, H. (1981). Some facts about Kurt Gödel. *Journal of Symbolic Logic 46*(3), 653–659.

Weyl, H. (1918). *Das Kontinuum.* Verlag von Veit and Comp., Leipzig. English translation Weyl (1994).

Weyl, H. (1994). *The Continuum.* Dover Publications, Inc., New York. Translated from the German by Stephen Pollard and Thomas Bole, with a foreword by John Archibald Wheeler and an introduction by Pollard. Corrected reprint of the 1987 translation [Thomas Jefferson Univ. Press, Kirksville, MO].

Whitehead, A. N. and B. Russell (1910). *Principia Mathematica.* Cambridge: Cambridge University Press. 3 vols. 1910, 1912, 1913.

Zermelo, E. (1904). Beweis dass jede Menge wohlgeordnet werden kann. *Mathematische Annalen 59*, 514–516. English translation in van Heijenoort (1967), pp. 139–141.

Zermelo, E. (1908). Untersuchungen über die Grundlagen der Mengenlehre I. *Mathematische Annalen 65*, 261–281. English translation in van Heijenoort (1967), pp. 200–215.

Index